# Agricultural Schedules and Federal Census Schedules 1850-1880

## Brecknock Township

## Lancaster County, Pennsylvania

*By James E. Frey*

***Agricultural Schedules and Federal Census Schedules 1850-1880***



Library of Congress Number: 97-61090

International Standard Book Number: 1-883294-53-3

*Published by*
Masthof Press
*R. R. #1, Box 20, Mill Road*
*Morgantown, PA 19543-9701*

# CONTENTS

# Foreword

Perhaps we should address readers of this book in Pennsylvania German. That is what the people of Brecknock Township spoke in 1850 through 1880. But that has changed by now and most of my dear readers would not be able to read German anymore. Only a few of the older persons in Brecknock use it today.

One of my Amish friends once said, "Ich kumm dich ewwi nett noch." He said, "I do not catch up with you; you get a year older each year." Former President Dwight D. Eisenhower once said, "Either you die young or you get old."

Methuselah reached the age of 969 years—almost a millenium. That is difficult to believe. But it is in the Bible, and I do believe it. Enoch walked with God, and God took him to be with Him. God loves company. Enoch pleased God, and God took him to be with Him. God so loved the world that He gave His only begotten Son, that whosoever believes in Him might have everlasting life and be forever with God. God wants our company. He wants us to be with Him.

Most of us missed the census years covered in this book. I did, but my parents both made it. It is so nice to have this record together with the record of the ages of the persons at the time.

Even the census taker did not always know what he was doing. When he reviewed his records, he said, "Perhaps this should go with the next name." Nevertheless, he did a good piece of work. We receive much more information as a result.

And now with the addition of the every-name index, this census record becomes doubly useful. And then there are parenthetic notations telling who lived with whom at the time, and who was this person's mother-in-law or grandfather or other relative—all helpful notations!

With all the work that has been done by others, this should be a real help to descendants and researchers. The census taker recorded the names used by the members of the family, but that should present no serious problems. We have the dates and ages.

The census taker, for example, may have written the name "Kitty" as received from the family, whereas the tombstone inscription records her name as "Catherine."

These census records are most important to us in that they are "official" and have the stamp of the United States government. We can rely on them. They are not rumor or hearsay. They tell it as it was in those years. I am so glad that they are accompanied by relatively poor maps showing where the families lived. And they are uniform; they tell the same about everyone.

- *Noah G. Good*
*Lancaster, Pennsylvania*
*June 1997*

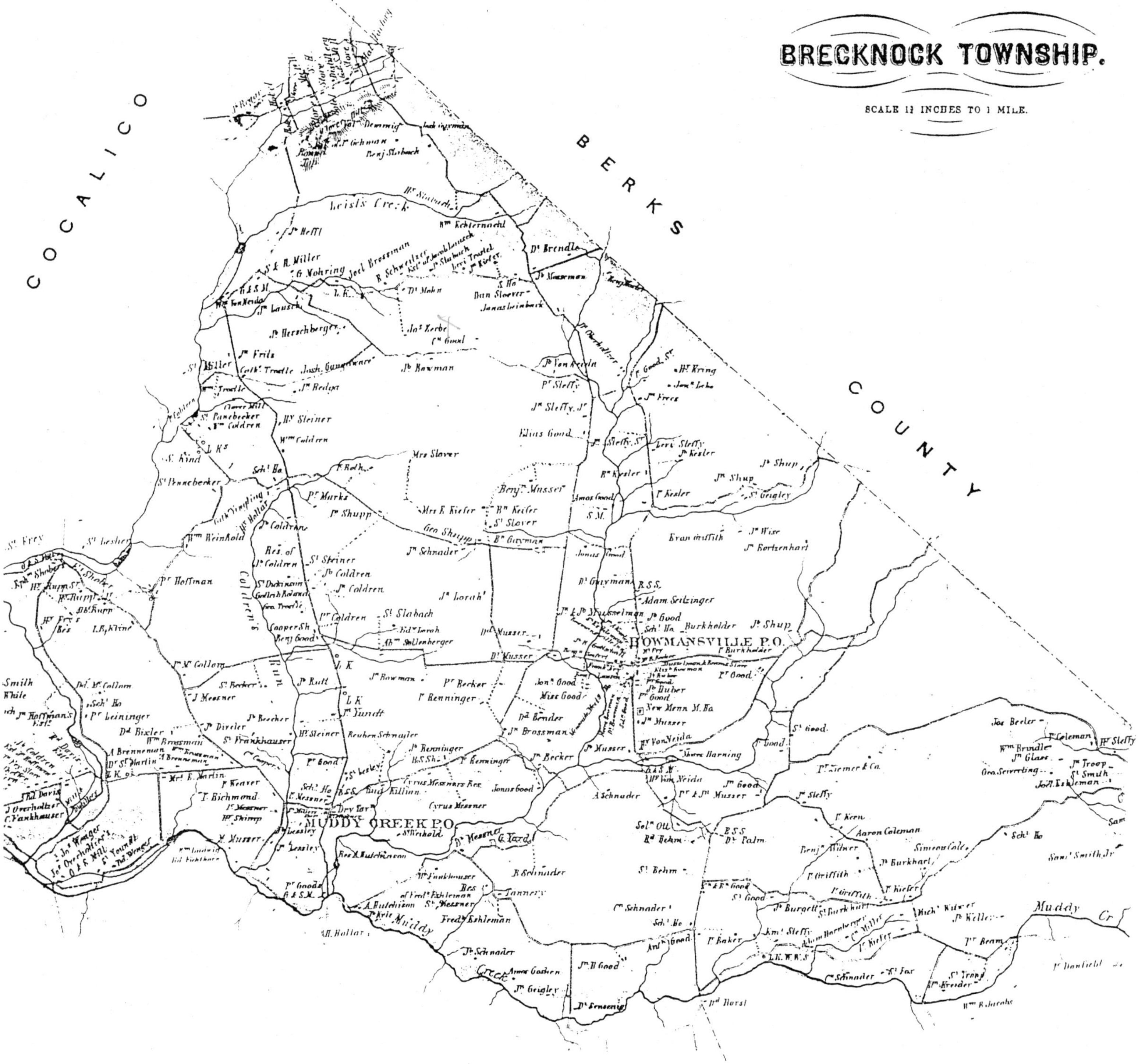


Brecknock Township in 1865

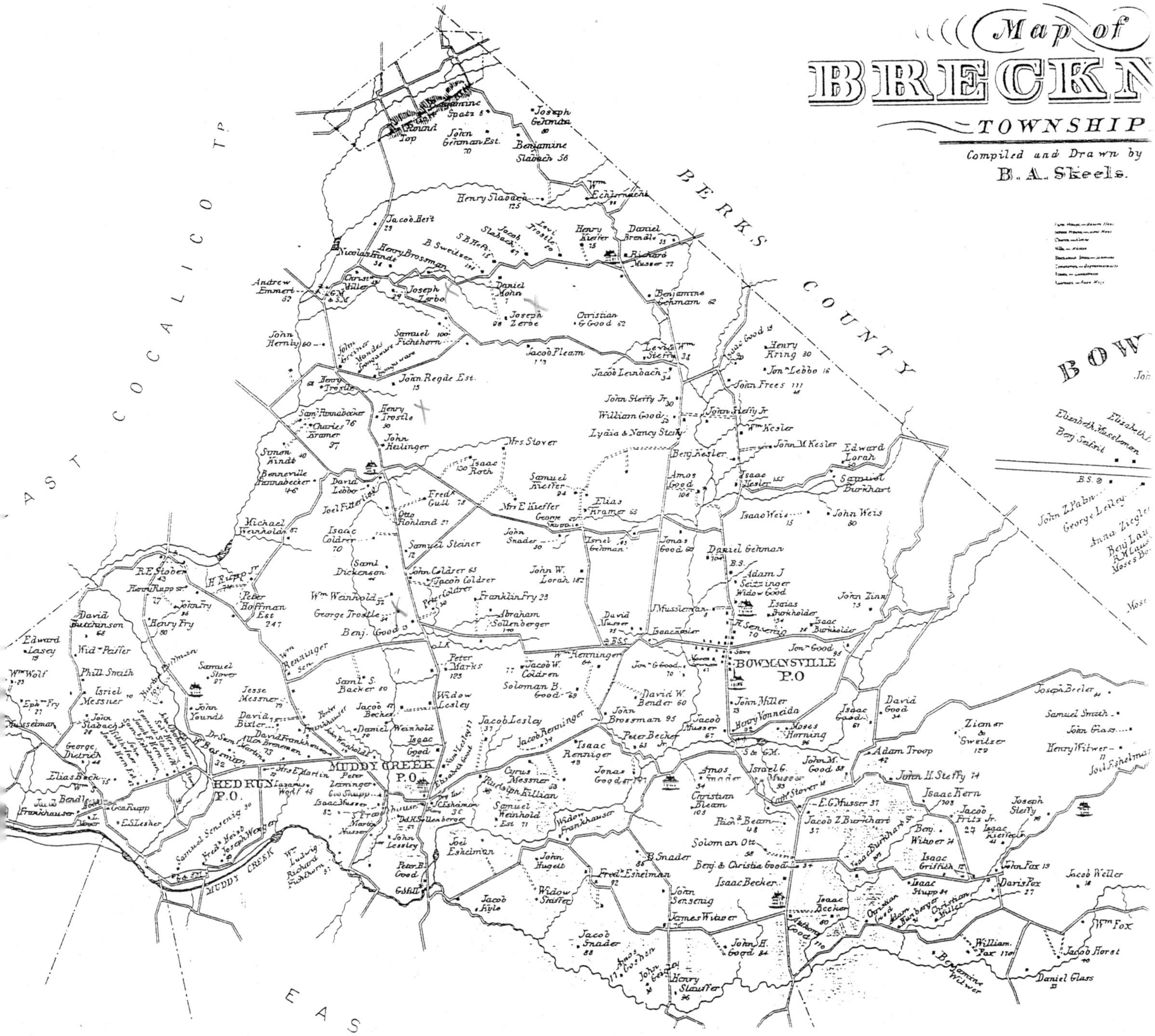


# Brecknock Township in 1875

# Brecknock Township in 1899

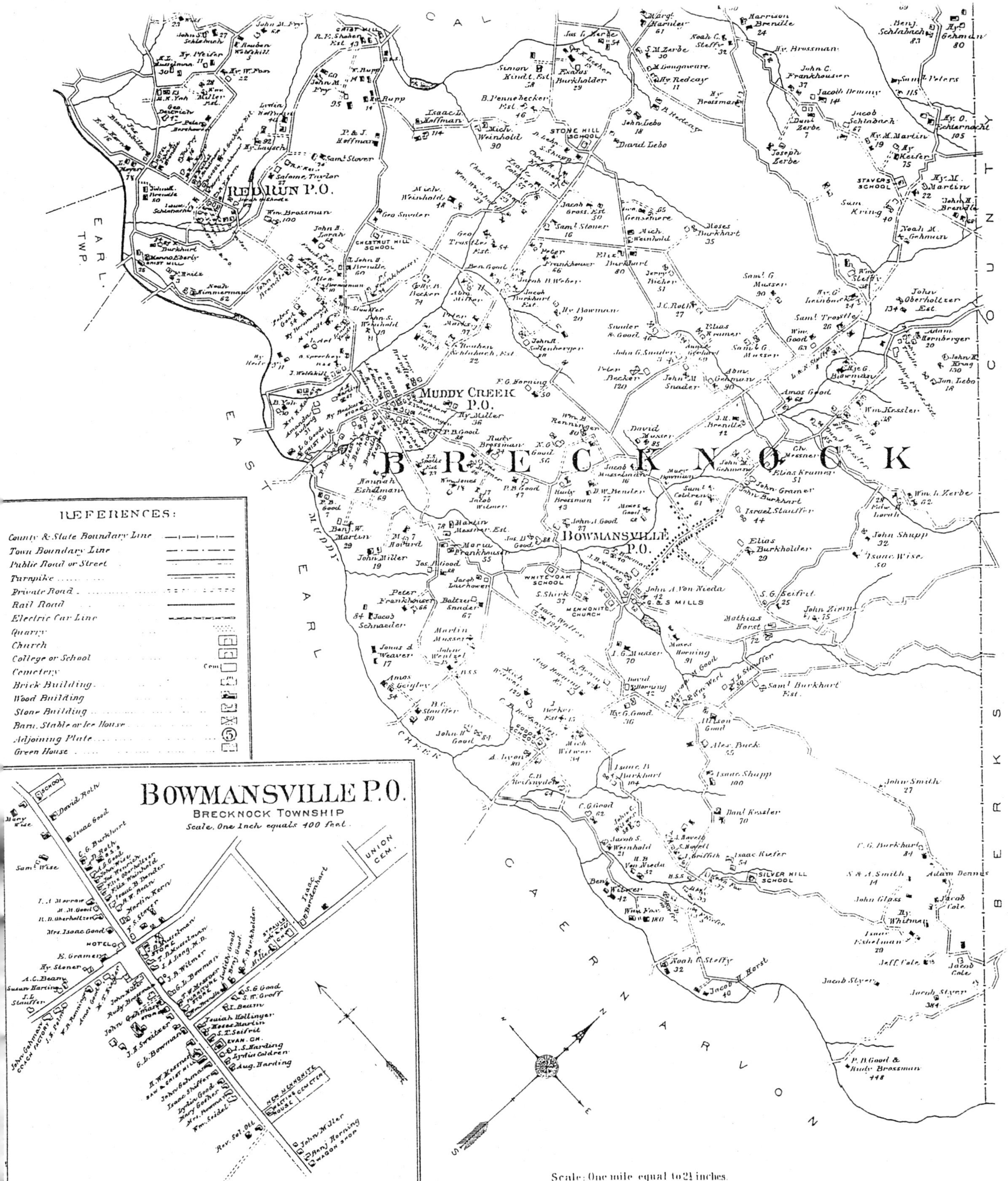


# Brecknock Township in 1899

# Introduction

There has been much emphasis placed on the United States Federal Census taken every ten years since 1790 for the purpose of Congressional district reapportionments, while other schedules undertaken by the Census Office have largely been ignored.

Between the years of 1850 and 1880, information was also gathered on agriculture, industry, mortality, and social data as **non-population schedules**. The enumerator (census taker) was required to gather accurate information; but, when it was not available, the farmer was then required to give a reasonable estimate. Household production was considered a part of farm production, but cabbage, potato patches, and family vegetable gardens were not counted.

> "For the purposes of the census, a farm was a managerial unit; whether owned or leased; whether contiguous or not; and whether cultivated, lying fallow, a wood lot, or sheep pasture. The property was considered one farm if all of it was under the care of the same person. Only farms with an annual produce valued at $100 were enumerated. In 1870 and 1880 no farm of less than three acres was reported unless $500 worth of produce from it had actually been sold during the year."[1]

Orchards and nurseries were considered to be a specialized form of agrarian units and were included in the agricultural schedules.

On or as close to June 1, the census taker traveled throughout the district assigned to him. Sometimes the same person took the decennial census also.

In 1850 and 1860, the **Agricultural Census Schedules** were divided into the following categories:

Name of owner, agent, or manager of farm
Improved acres of land
Cash value of farm
Value of farming implements and machinery
Livestock, June 1, 1850, or June 1, 1860
- Horses
- Asses and mules
- Milch cows
- Working oxen
- Other cattle

---

[1] Carmen Doune, *Federal Census Schedules, 1850-1880.* Reference paper No. 67. (Washington, D.C.: GSA, 1973), 18.

Sheep
Swine
Value of livestock
Produce during the year ending June 1, 1850 (1860)
Bushels of wheat
Bushels of rye
Bushels of Indian corn
Bushels of oats
Pounds of rice
Pounds of tobacco
Bales of ginned cotton, 400 pounds each
Pounds of wool
Bushels of peas and beans
Bushels of Irish potatoes
Bushels of sweet potatoes
Bushels of barley
Bushels of buckwheat
Value of orchard products in dollars
Gallons of wine
Value of produce of market gardens
Pounds of butter
Pounds of cheese
Tons of hay
Bushels of clover seed
Bushels of grass seed
Pounds of hops
Tons of drew rotted hemp
Tons of water rotted hemp
Other prepared hemp
Pounds of flax
Bushels of flaxseed
Pounds of silk cocoons
Pounds of maple sugar
Hogsheads of 100 pounds each of cane sugar
Gallons of molasses, and from what made
Pounds of beeswax
Pounds of honey
Value of homemade manufactures
Value animals slaughtered[2]

When the census taker for June 1870 collected his data, he had some changes to be recorded. Farm ownership had changed, either by inheritance, by sale, or by a division of the original farm into smaller units. Some of the categories were either expanded or combined, and other new categories were added. Hemp tonnage was combined while unimproved acreage was divided into woodland and other unimproved land. Wheat production was separated into spring harvests and winter harvests. New categories included in the 1870 census were the total amount of wages paid during the year, including the value of

[2] Washington, D.C., National Archives. Pennsylvania, Agricultural Census, Lancaster Co., Brecknock Twp., Record Group T1138, r5, r15.

board, gallons of milk sold, the dollar value of forest products, and the estimated value of all farm productions, including betterments and additions to stock.[3]

The government revised the 1870 agricultural census questions for 1880 and expanded it to 100 different questions to be answered. Those questions were divided into the following categories and sub-categories:

I. The name of the person who conducts the farm
II. Occupier is the owner of the farm
III. Occupier rents at a fixed money rental
IV. Occupier conducts farm for shares of products
V. Acres of land
- A. Improved acreage
  - 1. Tilled, including fallow and grass in rotation whether pasture or meadow
  - 2. Permanent meadows, pastures, orchards, vineyards
- B. Unimproved acreage
  - 1. Woodland and forest
  - 2. Other unimproved, including "old fields" not growing wood

VI. Farm values
- A. Dollar value of farm, including land, fences, and buildings
- B. Dollar value of farming implements and machines
- C. Dollar value of livestock

VII. Fences, cost of building and repairing fences in 1879
VIII. Cost of fertilizers purchased in 1879
IX. Labor
- A. Amount paid for wages for farm labor during 1879, including value of board
- B. Weeks hired labor in 1879 upon farm (and dairy) excluding housework

X. Estimated value of all farm productions (sold, consumed, or on hand for 1879)
XI. Grass lands
- A. Acreage-1879
  - 1. Mown
  - 2. Now mown
- B. Products harvested in 1879
  - 1. Hay
  - 2. Clover seed
  - 3. Grass seed

XII. Horses of all ages on hand June 1, 1880
XIII. Mules and asses of all ages on hand June 1, 1880
XIV. Neat cattle and their products
- A. Working oxen on hand June 1, 1880
- B. Milch cows on hand June 1, 1880
- C. Other cattle on hand June 1, 1880
- D. Calves dropped—1879
- E. Movement in 1879 for cattle of all ages
  - 1. Purchased
  - 2. Sold living
  - 3. Slaughtered
  - 4. Died, strayed, stolen and not recovered
- F. Gallons of milk sold, or sent to butter and cheese factories in 1879

[3] Ag census microfilm Record Group T1138, r26.

G. Pounds of butter made on the farm in 1879
H. Pounds of cheese made on the farm in 1879

XV. Sheep
- A. Number on hand June 1, 1880
- B. Number of lambs dropped
- C. Sheep and lambs
  1. Purchased
  2. Sold living
  3. Slaughtered
  4. Killed by dogs
  5. Died of disease
  6. Died of stress of weather
- D. Clip, spring, 1880, shorn and to be shorn
  1. Fleeces
  2. Weight in pounds

XVI. Swine on hand June 1, 1880

XVII. Poultry on hand, June 1, 1880, exclusive of spring hatching
- A. Barnyard
- B. Other

XVIII. Dozens of eggs produced in 1879

XIX. Cereal production in 1879
- A. Barley
  1. Acres planted
  2. Bushels yielded
- B. Buckwheat
  1. Acres planted
  2. Bushels yielded
- C. Indian corn
  1. Acres planted
  2. Bushels yielded
- D. Oats
  1. Acres planted
  2. Bushels yielded
- E. Rye
  1. Acres planted
  2. Bushels yielded
- F. Wheat
  1. Acres planted
  2. Bushels yielded

XX. Pulse
- A. Canada peas (dry) in bushels in 1879
- B. Beans, dry bushels in 1879

XXI. Fiber
- A. Flax area given in acres, 1879
- B. Bushels of flaxseed, 1879
- C. Tons of flax straw, 1879
- D. Pounds of flax fiber, 1879
- E. Acres of hemp
- F. Tons of hemp

XXII. Sugar
- A. Sorghum area given in acres, 1879
- B. Pounds of sorghum sugar, 1879
- C. Gallons of sorghum molasses, 1879
- E. Gallons of maple molasses, 1879

XXIII. Broom corn, 1879
- A. Acres
- B. Pounds harvested

XIV. Hops, 1879
- A. Acres
- B. Pounds harvested

XXV. Potatoes (Irish), 1879
- A. Acres planted
- B. Bushels

XXVI. Potatoes (sweet), 1879
- A. Acres planted
- B. Bushels harvested

XXVII. Tobacco, 1879
- A. Acres planted
- B. Pounds harvested

XXVIII. Orchards, 1879
- A. Apple production
    - 1. Acres
    - 2. Number of bearing trees
    - 3. Bushels harvested, 1879
- B. Peach production
    - 1. Acres
    - 2. Number of bearing trees
    - 3. Bushels harvested, 1879
- C. Total value of orchard products of all kinds sold or consumed

XXIX. Nurseries
- A. Acres
- B. Dollar value of produce sold in 1879

XXX. Vineyards
- A. Acres
- B. Pounds of grapes sold in 1879
- C. Gallons of wine made in 1879

XXXI. Market gardens, value of produce sold, 1879

XXXII. Bees, 1879
- A. Pounds of honey
- B. Pounds of beeswax

XXXIII. Forest products
- A. Cords of wood cut in 1879
- B. Dollar value of all forest products sold or consumed in 1879[4]

The records in this book were excerpted from the Agricultural Census Schedules found in the Lancaster County Historical Society, and cross-referenced with the particular year's Federal Census to pair up the likeliest farmer, using at least three criteria: farmer's name, acreage of farm, and value of land. Surnames are omitted in federal census notations when spelled the same as the farmer's name in the bold heading. If different, those surnames and varied spellings are included.

Readers can follow a particular ancestor as he farmed, his changing methods, his changing family, this changing to more scientific methods of farming and possibly into the next generation as the farm was often passed on to an offspring.

---

[4] Ag census microfilm Record Group T1138, r47.

# Brecknock Township—1850

### Samuel Bowman

- 25 acres improved and 20 acres unimproved land valued at $2100.
- 50 bu. wheat, 75 bu. Indian corn, 80 bu. oats, 50 bu. Irish potatoes, 3 bu. sweet potatoes, 150 lbs. butter, and 5 tons hay.
- Animals slaughtered valued at $60.

*Federal census lists Samuel, 60, a merchant; Elizabeth, 52. No property value given.*

### Jonas Musselman

- 80 acres of improved and 10 acres of unimproved land valued at $3000.
- Farm machinery valued at $85.
- 1 horse, 4 milch cows, 1 other cattle, 8 swine, valued at $200.
- 40 bu. wheat, 75 bu. rye, 150 bu. Indian corn, 150 bu. oats, 50 bu. Irish potatoes, 250 lbs. butter, 20 tons hay, 3 bu. clover seed.
- Animals slaughtered valued at $80.

*Federal census lists Jonas, 42; Levina, 34; Lucy, 16; Israel, 14; Eliza, 8. No property value given.*

### John Good, miller

- 30 acres improved and 50 acres unimproved land valued at $2500.
- Farm machinery valued at $75.
- 2 horses, 4 milch cows, 4 other cattle, 9 swine, valued at $220.
- 125 bu. rye, 50 bu. oats, 45 bu. Irish potatoes, 250 lbs. butter, 15 tons hay.
- Animals slaughtered valued at $85.

*Federal census names John, 45, a miller; Nancy, 41; Elizabeth, 18; Ephraim, 15, a farmer; and Nathan, 2, and property valued at $7000.*

### Daniel Bowman

- 60 acres improved and 45 acres unimproved land valued $4000.
- Farm machinery valued at $80.
- 2 horses, 6 milch cows, 3 other cattle, 7 swine, valued at $325.
- 100 bu. wheat, 50 bu. rye, 250 bu. Indian corn, 200 bu. oats, 50 bu. Irish potatoes, 450 lbs. butter, 18 tons hay, 6 bu. clover seed.
- Animals slaughtered valued at $70.

*Federal census lists Daniel, 53; Elizabeth, 52; Leah, 17; and property valued at $4000.*

### John Musselman

- 80 acres improved and 36 acres unimproved land valued at $4000.
- Farm machinery valued at $100.
- 5 horses, 5 milch cows, 9 other cattle, 14 swine, valued at $340.
- 85 bu. wheat, 80 bu. rye, 200 bu. Indian corn, 200 bu. oats, 125 bu. Irish potatoes, 350 lbs. butter, 22 tons hay, 2 bu. clover seed, 10 lbs. flax, 2 bu. flaxseed.
- Animals slaughtered valued at $114.

*Federal census does not have John listed.*

### Ulrich Burkholder

- 25 acres improved and 6 acres unimproved land valued at $1300.
- Farm machinery valued at $32.
- 1 horse, 2 milch cows, 3 swine, valued at $69.
- 35 bu. wheat, 10 bu., rye, 35 bu. Indian corn, 100 bu. oats, 40 bu. Irish potatoes, 100 lbs. butter, 4 tons hay, 2 bu. clover seed.
- Animals slaughtered valued at $67.

*Federal census lists Ulrich, 51; Esaias, 20, a laborer; Leah, 18; and property valued at $1300.*

### Nicholas Wise

- 50 acres improved and 25 acres unimproved land valued at $1300.
- Farm machinery valued at $75.
- 3 horses, 4 milch cows, 1 other cattle, 2 sheep, 3 swine, valued at $150.
- 30 bu. wheat, 30 bu. rye, 40 bu. Indian corn, 40 bu. oats, 30 bu. Irish potatoes, 100 lbs. butter, 6 tons hay.
- Animals slaughtered valued at $59.

*Federal census lists Nicholas, 72; Elizabeth, 71; Elizabeth, 20; and property valued at $1300.*

### Jacob Shup

- 40 acres improved and 40 acres unimproved land valued at $1200.
- Farm machinery valued at $50.
- 2 horses, 4 milch cows, 1 other cattle, 10 sheep, 6 swine, valued at $125.
- 12 bu. wheat, 40 bu. rye, 40 bu. Indian corn, 25 bu. oats, 1 bu. peas and beans, 30 bu. Irish potatoes, 40 lbs. butter, 5 tons hay, 1 bu. clover seed, 5 lbs. flax, 1 bu. flaxseed.
- Animals slaughtered valued at $63.

*Federal census lists Jacob, 48; Sophia, 38; Lydia, 18; Isaac, 14; Peter, 12; and property valued at $1200.*

### Elias Steffy

- 80 acres improved and 15 acres unimproved land valued at $1500.
- Farm machinery valued at $100.
- 4 horses, 6 milch cows, 3 other cattle, 9 swine, valued at $350.
- 65 bu. wheat, 7 bu. rye, 60 bu. Indian corn, 375 bu. oats, 40 bu. Irish potatoes, 400 lbs. butter, 15 tons hay, 1 bu. clover seed, 1 bu. other grass seed.
- Animals slaughtered valued at $100.

*Federal census lists Elias, 47; Elizabeth, 40; Noel, 21, a laborer; Eliza, 19; Jeramiah, 16, a laborer; John, 11; Levina, 9; and property valued at $1300.*

### Henry Schlebach

- 70 acres improved and 30 acres unimproved land valued at $2000.
- Farm machinery worth $500.
- 6 horses, 6 milch cows, 8 other cattle, 15 sheep, 6 swine, valued at $500.
- 60 bu. wheat, 30 bu. rye, 200 bu. Indian corn; 470 bu. rye, 100 bu. Irish potatoes, 100 lbs. butter, 40 tons hay, 2 bu. clover seed, 17 lbs. flax, 3 bu. flaxseed.
- Animals slaughtered valued at $100.

*Federal census lists Henry as Henry Slebach, 55; Lydia, 49; Jacob, 29, a cooper; Henry, 27, a laborer; Benjamin, 25, a laborer; Mary, 19; Lydia, 17; Samuel, 11; John, 7; and property valued at $2000.*

### John Gehman

- 50 acres improved and 43 acres unimproved land valued at $3000.
- Farm machinery valued at $50.
- 4 horses, 6 milch cows, 3 other cattle, 5 swine, valued at $225.
- 50 bu. wheat, 36 bu. rye, 80 bu. Indian corn, 75 bu. oats, 100 bu. Irish potatoes, 200 lbs. butter, 12 tons hay, 2 bu. clover seed, 2 bu. other grass seed, 16 lbs. flax, 3 bu. flaxseed.
- Animals slaughtered valued at $83.

*Federal census lists John, 29; Lydia, 28; Joshua, 3; Ephriam, 1; Susanna, 1/12; William Steiner, 14. No property value given.*

### Joseph Gehman

- 88 acres improved and 20 acres unimproved land valued at $3000.
- Farm machinery valued at $100.
- 3 horses, 3 milch cows, 9 other cattle, 7 sheep, 10 swine, valued at $207.
- 100 bu. wheat, 25 bu. rye, 180 bu. Indian corn, 200 bu. oats, 80 bu. Irish potatoes, 250 lbs.; butter, 24 tons hay, 1 bu. clover seed, 6 lbs. flax, 1 bu. flaxseed.
- Animals slaughtered valued at $85.

*Federal census lists Joseph, 40; Esther, 35; Isaac, 14; Anna, 11; Henry, 6; and property valued at $3000.*

### Christian Musser

- 60 acres improved and 5 acres unimproved land valued at $1000.
- Farm machinery valued at $50.
- 2 horses, 4 milch cows, 2 other cattle, 4 sheep, 5 swine, valued at $148.
- 50 bu. wheat, 10 bu. rye, 50 bu. Indian corn, 100 bu. oats, 40 bu. Irish potatoes, 100 lbs. butter, 15 tons hay, 2 bu. clover seed, 1 bu. other grass seed, 5 lbs. flax, 1 bu. flaxseed.
- Animals slaughtered valued at $45.

*Federal census lists Christian, 34; Julian, 17; Richard, 15; Daniel, 14. No property value given.*

### John Kiefer

- 30 acres improved and 18 acres unimproved land valued at $1200.
- Farm machinery valued at $50.
- 1 horse, 5 milch cows, 3 swine valued at $100.
- 20 bu. wheat, 35 bu. rye, 75 bu. Indian corn, 100 bu. oats, 20 bu. Irish potatoes, 150 lbs. butter, 6 tons hay.
- Animals slaughtered valued at $55.

*Federal census lists John, 39; Catherine, 35; Caroline, 17, Catherine, 10; Henry, 9; and property valued at $1200.*

### Isaac Coldren
• 60 acres improved and 6 acres unimproved land valued at $2000.
• Farm machinery valued at $70.
• 2 horses, 4 milch cows, 4 other cattle, 11 sheep, 4 swine, valued at $226.
• 30 bu. wheat, 30 bu. rye, 100 bu. Indian corn, 100 bu. oats, 60 bu. Irish potatoes, 200 lbs. butter, 5 tons hay.
• Animals slaughtered valued at $73.
*Federal census lists Isaac, 45; Mary, 47; Samuel, 16, a laborer; Elizabeth, 14; Jesse, 12; Harrison, 10; Mary, 8; and property valued at $2000.*

### Daniel Hertz
• 40 acres improved and 7 acres unimproved land valued at $1800.
• Farm machinery valued at $75.
• 2 horses, 2 milch cows, 2 other cattle, 3 swine, valued at $100.
• 70 bu. wheat, 35 bu. Indian corn, 10 bu. oats, 75 bu. Irish potatoes, 80 lbs. butter, 5 tons hay, 1 bu. other grass seed.
• Animals slaughtered valued at $30.
*Federal census lists Daniel, 34; Elizabeth, 34; Levina Merks, 11; and property valued at $1800.*

### Peter Yohn
• 100 acres improved and 56 acres unimproved land valued at $6000.
• Farm machinery valued at $100.
• 4 horses, 7 milch cows, 14 other cattle, 2 sheep, 8 swine, valued at $505.
• 250 bu. wheat, 30 bu. rye, 350 bu. Indian corn, 200 bu. oats, 75 bu. Irish potatoes, 300 lbs. butter, 40 tons hay, 9 bu. clover seed, 5 bu. other grass seed.
• Animals slaughtered valued at $105.
*Federal census lists Peter, 63; Catherine, 59; Rachel, 26; Ellen, 20; and property valued at $6000.*

### John Steiner
• 57 acres improved and 9 acres unimproved land valued at $2500.
• Farm machinery valued at $50.
• 2 horses, 3 milch cows, 6 other cattle, 7 swine, valued at $211.
• 100 bu wheat, 25 bu. rye, 100 bu. Indian corn, 150 bu. oats, 50 bu. Irish potatoes, 6 bu. buckwheat, 200 lbs. butter, 20 tons hay, 1 bu. clover seed, 1 bu. other grass seed, 12 lbs. flax, 2 bu. flaxseed.
• Animals slaughtered valued at $78.
*Federal census lists John, 53, a mason; Judith, 51; Mary, 26; Elizabeth, 23; Andrew, 21, a miller; Catherine, 17; Reuben, 15, a laborer; Ephraim, 13; Edward, 12; and property valued at $2500.*

### Anthony Good
• 63 acres improved and 45 acres unimproved land valued at $2500.
• Farm machinery valued at $100.
• 4 horses, 3 milch cows, 8 other cattle, 4 swine, valued at $400.
• 45 bu. wheat, 8 bu. rye, 70 bu. Indian corn, 300 bu. oats, 7 tons hay, 6 bu. clover seed.
*Federal census lists Anthony, 33; Catherine, 22; Isaac Zimmerman, 20, a laborer; and property valued at $2500.*

### John H. Good
• 60 acres improved and 20 acres unimproved land valued at $2500.
• Farm machinery valued at $150.
• 3 horses, 3 milch cows, 6 other cattle, 2 sheep, 6 swine, valued at $300.
• 50 bu. wheat, 30 bu. rye, 100 bu. Indian corn, 150 bu. oats, 40 bu. Irish potatoes, 200 lbs. butter, 15 tons hay, 1 bu. other grass seed.
• Animals slaughtered valued at $70.
*Federal census lists John, 35; Levina, 25; Catherine, 5; Holly Ann, 3; Martin, 11/12; John Kuch, 16, a laborer; and property valued at $7000.*

### Daniel Sensenig
• 112 acres improved and 3 acres unimproved land valued at $6000.
• Farm machinery valued at $300.
• 3 horses, 4 milch cows, 7 other cattle, 5 swine, valued at $300.
• 140 bu. wheat, 50 bu. rye, 300 bu. Indian corn, 300 bu. oats, 60 bu. Irish potatoes, 200 lbs. butter, 20 tons hay, produce of market gardens valued at $5.
• Animals slaughtered valued at $200.
*Federal census lists Daniel, 38; Elizabeth, 37; Ann, 12; Jacob, 10, Amos, 8; Isaac, 4; Matilda, 3; John, 6/12; and property valued at $6000.*

## Christian Schneder
- 30 acres improved and 11 acres unimproved land valued at $2000.
- Farm machinery valued at $30.
- 1 horse, 3 milch cows, 7 other cattle, 2 swine, valued at $100.
- 70 bu. wheat, 100 bu. Indian corn, 467 bu. oats, 30 bu. Irish potatoes, 7 bu. buckwheat, 200 lbs. butter, 7 tons hay.
- Animals slaughtered valued at $65.

*Federal census lists Christian, 34; Elizabeth, 36; Sarah, 9; Caroline, 7; Barton, 6; Harrison, 5; Christianna, 3; Catherine, 1; and property valued at $2000.*

## Jacob Messner
- 78 acres improved and 12 acres improved land valued at $3000.
- Farm machinery valued at $100.
- 4 horses, 3 milch cows, 7 other cattle, 6 swine, valued at $400.
- 40 bu. wheat, 10 bu. rye, 100 bu. Indian corn, 200 bu. oats, 20 bu. Irish potatoes, 100 lbs. butter, 6 bu. clover seed.
- Animals slaughtered valued at $70.

*Federal census lists Jacob, 41; Sarah, 34; John, 14; Elizabeth, 10; Jacob Burkhart, 23, a laborer; Jacob Simes(?), 7; and property valued at $3000.*

## William Schneder
- 82 acres improved and 30 acres unimproved land valued at $4000.
- Farm machinery valued at $100.
- 7 horses, 6 milch cows, 8 other cattle, 7 swine, valued at $356.
- 100 bu. wheat, 20 bu. rye, 150 bu. Indian corn, 150 bu. oats, 40 bu. Irish potatoes, 200 lbs. butter, 25 tons hay, 1 bu. clover seed, 1 bu. other grass seed.
- Animals slaughtered valued at $53.

*Federal census lists William, 50; Nelly, 46; Louisa, 22; Jacob, 19, a laborer; William, 17, a laborer; Eliza Weitzel, 10; and property valued at $4000.*

## Samuel Beam
- 85 acres improved and 35 acres unimproved land valued at $4500.
- Farm machinery valued at $100.
- 4 horses, 6 milch cows, 13 sheep, 6 swine, valued at $374.
- 50 bu. wheat, 8 bu. rye, 300 bu. Indian corn, 300 bu. oats, 30 bu. Irish potatoes, 400 lbs. butter, 30 tons hay, 5 bu. clover seed, 2 bu. other grass seed.
- Animals slaughtered valued at $175.

*Federal census lists Samuel, 43; Abraham Weller, 63, a laborer; Catherine Stover, 28; and property valued at $2500.*

## Abraham Good
- 29 acres improved and 6 acres unimproved land valued at $1500.
- Farm machinery valued at $30.
- 1 horse, 3 milch cows, 2 other cattle, 4 swine, valued at $105.
- 50 bu. wheat, 10 bu. rye, 30 bu. Indian corn, 40 bu. oats, 25 bu. Irish potatoes, 100 lbs. butter, 5 tons hay, 2 bu. clover seed.
- Animals slaughtered valued at $49.

*Federal census lists Abraham, 57; Elizabeth, 19; Benjamin, 18, a laborer; and property valued at $1500.*

## Isaac Becker
- 50 acres improved and 14 acres unimproved land valued at $2500.
- Farm machinery valued at $150.
- 2 horses, 3 milch cows, 8 other cattle, 4 sheep, 6 swine, valued at $200.
- 50 bu. wheat, 40 bu. rye, 125 bu. Indian corn, 100 bu. oats, 50 bu. Irish potatoes, 300 lbs. butter, 6 tons hay, 3 bu. clover seed.
- Animals slaughtered valued at $65.

*Federal census lists Isaac, 36; Sarah, 33, Caroline, 9; Susanna, 7; John, 3; and property valued at $2500.*

## Emanuel Steffy
- 55 acres improved and 7 acres unimproved land valued at $2500.
- Farm machinery valued at $200.
- 3 horses, 5 milch cows, 5 other cattle, 6 sheep, 4 swine, valued at $500.
- 100 bu. wheat, 70 bu. rye, 100 bu. Indian corn, 100 bu. oats, 20 bu. Irish potatoes, 300 lbs. butter, 10 tons hay.
- Animals slaughtered valued at $70.

*Federal census lists Emanuel, 53; Barbara, 52; Jacob, 23, a cooper; Polly, 18; Samuel, 15, a laborer; and property valued at $2300.*

### Samuel Fox
- 70 acres improved and 20 acres unimproved land valued at $1800.
- Farm machinery valued at $100.
- 5 horses, 4 milch cows, 8 other cattle, 6 swine, valued at $250.
- 70 bu. wheat, 30 bu. rye, 100 bu. Indian corn, 150 bu. oats, 10 bu. Irish potatoes, 750 lbs. butter, 11 tons hay, 1 bu. clover seed.
- Animals slaughtered valued at $45.

*Federal census lists Samuel, 48; Elizabeth, 50; John, 21, a carpenter; Catherine, 18; Davis, 15, a laborer; William, 11; Henrietta, 17; Barbara, 7; and property valued at $1800.*

### Jacob Burkhart
- 74 acres improved and 25 acres unimproved land valued at $2500.
- Farm machinery valued at $79.
- 3 horses, 5 milch cows, 6 other cattle, 12 swine, valued at $249.
- 100 bu. wheat, 45 bu. rye, 100 bu. Indian corn, 225 bu. oats, 15 bu. Irish potatoes, 300 lbs. butter, 15 tons hay, 1 bu. clover seed.
- Animals slaughtered valued at $65.

*Federal census lists Jacob, 47; Anna, 38; Catherine, 17; Hetty, 10; Judith, 8; Susanna, 6; Barbara, 4; John Burkhart, 71; and property valued at $2500.*

### Samuel Good
- 80 acres improved and 40 acres unimproved land valued at $2700.
- Farm machinery valued at $90.
- 3 horses, 6 milch cows, 10 other cattle, 6 sheep, 6 swine, valued at $300.
- 95 bu. wheat, 50 bu. rye, 185 bu. Indian corn, 200 bu. oats, 90 bu. Irish potatoes, 2 bu. barley, 300 lbs. butter, 20 tons hay, 1 bu. clover seed, 1 bu. other grass seed.
- Animals slaughtered valued at $70.

*Federal census lists Samuel, 36, Menn. clergy; Esther, 39; Mary, 14; John, 11; Christian, 9; Susanna, 7; Jacob, 5; Barbara, 1; Daniel Good, 43, a weaver; and property valued at $2700.*

### Joseph Beler
- 50 acres improved and 30 acres unimproved land valued at $700.
- Farm machinery valued at $70.
- 2 horses, 4 milch cows, 4 other cattle, 12 sheep, 8 swine, valued at $200.
- 30 bu. wheat, 30 bu. rye, 70 bu. Indian corn, 70 bu. oats, 100 bu. Irish potatoes, 150 lbs. butter, 10 tons hay.
- Animals slaughtered valued at $65.

*Federal census lists Joseph, 49; Elizabeth, 51; Susanna, 23; Mary, 12; Levina, 10; Moses, 8; Joseph, 6; and property valued at $700.*

### Christian Schneder
- 38 acres improved and 11 acres unimproved land valued at $1800.
- Farm machinery valued at $50.
- 1 horse, 5 milch cows, 3 other cattle, 3 swine, valued at $110.
- 45 bu. wheat, 100 bu. Indian corn, 70 bu. oats, 40 bu. Irish potatoes, 500 lbs. butter, 10 tons hay, 2 bu. clover seed.
- Animals slaughtered valued at $37.

*Federal census lists Christian, 47; Margaret, 46; Susanna Schweigart, 7; and property valued at $1800.*

### Isaac Kern
- 65 acres improved and 35 acres unimproved land valued at $1500.
- Farm machinery valued at $75.
- 4 horses, 4 milch cows, 8 other cattle, 8 sheep, 9 swine, valued at $200.
- 100 bu. wheat, 15 bu. rye, 150 bu. Indian corn, 100 bu. oats, 40 bu. Irish potatoes, 250 lbs. butter, 15 tons hay, 2 bu. clover seed, 1 bu. other grass seed.
- Animals slaughtered valued at $70.

*Federal census lists Isaac, 39; Elizabeth, 28; Henry, 12; Caroline, 10; Barbara, 9; Catherine, 8; Elizabeth, 7; Martin, 6; Adam, 5; John, 4; Isaac, 1; and property valued at $1500.*

### Richard Beam
- 43 acres improved and 11 acres unimproved land valued at $1500.
- Farm machinery valued at $90.
- 2 horses, 3 milch cows, 5 other cattle, 3 swine, valued at $125.
- 30 bu. wheat, 40 bu. rye, 90 bu. Indian corn, 150 bu. oats, 25 bu. Irish potatoes, 200 lbs. butter, 10 tons hay, 2 bu. clover seed, 2 bu. other grass seed.
- Animals slaughtered valued at $55.

*Federal census lists Richard, 27; Sarah, 27; Cassy, 29; and property valued at $1300.*

## Adam Kessler
• 40 acres improved and 7 acres unimproved land valued at $2000.
• Farm machinery valued at $150.
• 3 horses, 4 milch cows, 6 other cattle, 9 sheep, 5 swine, valued at $200.
• 30 bu. wheat, 25 bu. rye, 50 bu. Indian corn, 50 bu. oats, 30 bu. Irish potatoes, 100 lbs. butter, 10 tons hay, 2 bu. clover seed, 18 lbs. flax, 3 bu. flaxseed.
• Animals slaughtered valued at $47.

*Federal census lists Adam, 49; Rachel, 43; Susanna, 23; Lydia, 22; Rebecca, 20; Sarah, 16; Cassy, 12; Nancy, 10; John, 5; Elizabeth, 2; and property valued at $2000.*

## Ephraim Shober
• 50 acres improved and 15 acres unimproved land valued at $4000.
• Farm machinery valued at $100.
• 2 horses, 4 milch cows, 1 other cattle, 6 swine, valued at $ 280.
• 80 bu. wheat, 150 bu. Indian corn, 80 bu. oats, 30 bu. Irish potatoes, 200 lbs. butter, 15 tons hay, 1 bu. clover seed.
• Animals slaughtered valued at $75.

*Federal census lists Ephraim, 52, a miller; Susanna, 48; Fay Ann, 23; Reuben, 22, a miller; Solomon Frey, 13; and property valued at $8000.*

## Isaac Fry
• 60 acres improved and 20 acres unimproved land valued at $2000.
• Farm machinery valued at $60.
• 2 horses, 3 milch cows, 3 other cattle, 4 sheep, 3 swine, valued at $225.
• 150 bu. wheat, 50 bu. rye, 200 bu. Indian corn, 350 bu. oats, 25 bu. Irish potatoes, 15 bu. barley, 100 lbs. butter, 18 tons hay, 3 bu. clover seed, 1 bu. other grass seed.
• Animals slaughtered valued at $40.

*Federal census lists Isaac, 43; Catherine, 40; William, 13; and property valued at $600.*

## John Peiffer
• 24 acres improved and 10 acres unimproved land valued at $1000.
• Farm machinery valued at $40.
• 2 horses, 3 milch cows, 7 swine, valued at $165.
• 25 bu. wheat, 12 bu. rye, 10 bu. Indian corn, 70 bu. oats, 14 bu. Irish potatoes, 100 lbs. butter, 5 tons hay.
• Animals slaughtered valued at $25.

*Federal census lists John, 56; Regina, 55; George, 16, a laborer; Henry Frederic, 7; and property valued at $1000.*

## William Schlebach
• 50 acres improved and 10 acres unimproved land valued at $2000.
• Farm machinery valued at $100.
• 3 horses, 4 milch cows, 7 other cattle, 4 swine, valued at $282.
• 60 bu. wheat, 75 bu. rye, 200 bu. Indian corn, 100 bu. oats, 23 bu. Irish potatoes, 200 lbs. butter, 12 tons hay, 1 bu. other grass seed.
• Animals slaughtered valued at $54.

*Federal census lists William Slebach, 40; Sabina Slebach, 44; John Slebach, 16, a laborer; Susanna Slebach, 14; Lydia Slebach, 13; Sarah Slebach, 10; William Slebach, 4; and property valued at $2000.*

## Jacob Kern
• 23 acres improved and 8 acres unimproved land valued at $1400.
• Farm machinery valued at $45.
• 1 horse, 3 milch cows, 3 other cattle, 6 swine, valued at $78.
• 40 bu. wheat, 40 bu. Indian corn, 100 bu. oats, 15 bu. Irish potatoes, 100 lbs. butter, 6 tons hay.
• Animals slaughtered valued at $38.

*Federal census lists Jacob, 38; Fayanne, 33; Solomon, 9; Edward, 8; Jacob, 4; Henry Adam, 3; Richard, 2; and property valued at $1400.*

## Jacob Spatz
• 40 acres improved and 40 acres unimproved land valued at $3000.
• Farm machinery valued at $200.
• 3 horses, 4 milch cows, 11 other cattle, 17 sheep, 5 swine, valued at $482.
• 104 bu. wheat, 150 bu. Indian corn, 200 bu. oats, 8 lbs. wool, 100 bu. Irish potatoes, 400 lbs. butter, 25 tons hay, 1 bu. other grass seed.
• Animals slaughtered valued at $75.

*Federal census lists Jacob, 51; Elizabeth, 42; Mary, 18; John, 16, a laborer; Jacob, 11; Elizabeth, 9; Melinda, 7; and property valued at $3000.*

### Abraham Trostel

- 100 acres improved and 26 acres unimproved land valued at $4200.
- Farm machinery valued at $100.
- 3 horses, 6 milch cows, 3 other cattle, 8 swine, valued at $253.
- 50 bu. wheat, 25 bu. rye, 100 bu. Indian corn, 170 bu. oats, 18 lbs. wool*, 40 bu. Irish potatoes, 400 lbs. butter, 15 tons hay, 1 bu. clover seed, 1 bu. other grass seed, 2 lbs. hops, 1 bu. flaxseed.
- Animals slaughtered valued at $80.

*Note: This appears to have been entered in the wrong line because the following entry has 14 sheep and no wool entered (See William Winehold below).

*Federal census lists Abraham, 65; Ann Mary, 63; Henry, 30, a laborer; Catherine, 27; Sarah, 23; John, 11; and property valued at $4200.*

### William Winehold

- 59 acres improved and 30 acres unimproved land valued at $1200.
- Farm machinery valued at $40.
- 3 horses, 3 milch cows, 4 other cattle, 14 sheep, 3 swine, valued at $149.
- 20 bu. wheat, 40 bu. rye, 30 bu. Indian corn, 75 bu. oats, (wool, see Abraham Trostel above), 40 bu. Irish potatoes, 200 lbs. butter, 12 tons hay, 20 lbs. flax, 3 bu. flaxseed.
- Animals slaughtered valued at $70.

*Federal census lists William, 52; Mary, 47; Peter, 21, a laborer; William, 18, a laborer; Michael, 17, a laborer; Rebecca, 15; Cassiah, 12 (female); John, 11; Levi, 9; Jacob, 2; and property valued at $1200.*

### Peter Hoffman

- 94 acres improved and 20 acres unimproved land valued at $2700.
- Farm machinery valued at $60.
- 2 horses, 6 milch cows, 4 other cattle, 8 sheep, 4 swine, valued at $204.
- 50 bu. wheat, 12 bu. rye, 40 bu. Indian corn, 100 bu. oats, 15 lbs. wool, 15 bu. Irish potatoes, 300 lbs. butter, 12 tons hay, 2 bu. other grass seed.
- Animals slaughtered valued at $51.

*Federal census lists Peter, 45; Christianna, 35; Lydia, 12; Caroline, 9; Israel, 4; Peter Hoshaar, 15, a laborer; and property valued at $2700.*

### Simon Delier

- 60 acres improved and 40 acres unimproved land valued at $2500.
- Farm machinery valued at $100.
- 3 horses, 5 milch cows, 3 other cattle, 3 sheep, 4 swine, valued at $227.
- 50 bu. rye, 40 bu. Indian corn, 130 bu. oats, 8 lbs. wool, 35 bu. Irish potatoes, 300 lbs. butter, 10 tons hay, 2 bu. clover seed, 15 lbs. flax, 2 bu. flaxseed.
- Homemade manufactures valued at $15.
- Animals slaughtered valued at $47.

*Federal census lists Simon, 77; Elizabeth, 73; Catherine, 47; Susanna, 37; John, 35, a laborer; and property valued at $2500.*

### Jacob Lessly

- 80 acres improved and 20 acres unimproved land valued at $3000.
- Farm machinery valued at $100.
- 4 horses, 4 milch cows, 6 other cattle, 4 sheep, 6 swine, valued at $500.
- 100 bu. wheat, 30 bu. rye, 200 bu. Indian corn, 250 bu. oats, 8 lbs. wool, 30 bu. Irish potatoes, 200 lbs. butter, 25 tons hay, 2 bu. clover seed.
- Animals slaughtered valued at $75.

*Federal census lists Jacob, 44; Hannah, 33; Samuel, 18, a laborer; Jacob, 11; John, 8; Henrietta, 6; Joseph, 4; Henry, 2; Sophia, 10/12; Susanna Litch, 14; and property valued at $3000.*

### Michael Martin

- 30 acres improved and 2 acres unimproved land valued at $1500.
- Farm machinery valued at $15.
- 1 horse, 2 milch cows, 1 other cattle, 2 swine, valued at $50.
- 35 bu. wheat, 9 bu. rye, 40 bu. Indian corn, 40 bu. oats, 8 bu. Irish potatoes, 100 lbs. butter, 6 tons hay.
- Animals slaughtered valued at $27.

*Federal census lists Michael, 68, a gunsmith; Susanna, 61; Mary, 22; and property valued at $1500.*

### Peter Martin

- 40 acres improved and 10 acres unimproved land valued at $4000.
- 3 horses, 2 milch cows, 4 swine, valued at $120.
- 30 bu. wheat, 30 bu. rye, 60 bu. Indian corn, 150 bu. oats, 20 bu. Irish potatoes, 15 lbs. butter, 4 tons hay, 1 bu. clover seed.

• Animals slaughtered valued at $35.

*Federal census lists Peter, 55, a physician; Elizabeth, 55; Elijah, 23, a farmer; Elizabeth, 18; Jesse, 11; and property valued at $4000.*

### Joseph Overholtzer

• 12 acres improved land valued at $2000.
• Farm machinery valued at $40.
• 1 horse, 2 milch cows, 1 other cattle, 5 swine, valued at $138.
• 62 bu. wheat, 20 bu. Indian corn, 20 bu. oats, 15 bu. Irish potatoes, 200 lbs. butter, 2 tons hay.
• Homemade manufactures valued at $4.
• Animals slaughtered valued at $80.

Note: The difference in property value could mean the mill was included in the federal census total.

*Federal census lists Joseph, 29, a miller; Magdalene, 21; Mary, 4; Daniel, 2; John Oberholtzer, 31, a miller; Promilia Fasnacht, 14; and property valued at $7000.*

### Christian Martin

• 106 acres improved and 25 acres unimproved land valued at $3400.
• Farm machinery valued at $162.
• 4 horses, 6 milch cows, 3 other cattle, 2 swine, valued at $450.
• 190 bu. wheat, 10 bu. rye, 200 bu. Indian corn, 400 bu. oats, 30 bu. Irish potatoes, 300 lbs. butter, 18 tons hay, 6 bu. clover seed.
• Animals slaughtered valued at $62.

*Federal census lists Christian, 26; Elizabeth, 29; Barbara, 4; Abraham, 3; Elizabeth, 4/12; Isaac Martin, 28, a farmer; Mary Newpher, 17; and property valued at $3400.*

### Richard Davis

• 85 acres improved and 25 acres unimproved land valued at $3500.
• Farm machinery valued at $40.
• 4 horses, 6 milch cows, 3 other cattle, 3 swine, valued at $265.
• 60 bu. wheat, 6 bu. rye, 125 bu. Indian corn, 150 bu. oats, 25 bu. Irish potatoes, 60 lbs. butter, 10 tons hay, 5 lbs. flax, 1 lb. flaxseed.
• Animals slaughtered valued at $32.

*Federal census lists Richard, 61; Catherine, 58; Richard, 25, a carpenter; Lydia, 24; Rees H., 22, a carpenter; Samuel B., 18, a laborer; and property valued at $3500.*

### John Brossman

• 60 acres improved and 10 acres unimproved land valued at $4000.
• Farm machinery valued at $125.
• 4 horses, 4 milch cows, 5 other cattle, 3 sheep, 3 swine, valued at $435.
• 100 bu. wheat, 50 bu. rye, 150 bu. Indian corn, 200 bu. oats, 7 lbs. wool, 25 bu. Irish potatoes, 6 bu. barley, 200 lbs. butter, 20 tons hay, 3 bu. clover seed, 3 bu. other grass seeds.
• Animals slaughtered valued at $55.

*Federal census lists John, 46; Hannah, 46; William, 18, a laborer; Elizabeth, 14; Rudolph, 9; Emanuel Arnold, 46, a laborer; Sarah Stanch, 32; and property valued at $4000.*

### John Fry

• 10 acres improved and 4 acres unimproved land valued at $1000.
• 2 milch cows, 2 swine, valued at $46.
• 75 bu. wheat, 60 bu. rye, 100 bu. Indian corn, 100 bu. oats, 50 bu. Irish potatoes, 200 lbs. butter, 15 tons hay, 24 lbs. flax, 2 bu. flaxseed.
• Homemade manufactures valued at at $10.
• Animals slaughtered valued at $80.

*Federal census lists John, 74; Rachel, 27; Mary Hoshaar, 9; and property valued at $1000.*

### John Hoffman

• 100 acres improved and 50 acres unimproved land valued at $2000.
• Farm machinery valued at $79.
• 4 horses, 4 milch cows, 6 other cattle, 4 swine, valued at $172.
• 50 bu. wheat, 25 bu. rye, 100 bu. Indian corn, 150 bu. oats, 35 bu. Irish potatoes, 1 bu. sweet potatoes, 8 bu. buckwheat, 100 lbs. butter, 25 tons hay, 10 lbs. flax, 1 bu. flaxseed.
• Homemade manufactures valued at $5.
• Animals slaughtered valued at $40.

*Federal census lists John, 54; Catherine, 55; Samuel, 20, a laborer; John, 18, a laborer; George, 15, laborer; Elizabeth Wolfkill; and property valued at $2800.*

### Henry Fry

- 97 acres improved and 35 acres unimproved land valued at $3700.
- Farm machinery valued at $150.
- 4 horses, 7 milch cows, 6 other cattle, 4 sheep, 3 swine, valued at $352.
- 70 bu. wheat, 80 bu. Indian corn, 12 lbs. wool, 18 bu. Irish potatoes, 300 lbs. butter, 6 tons hay.
- Animals slaughtered valued at $82.

*Federal census lists Henry, 35; Sarah, 29; Eliza, 7; Susanna, 6; Martin Lorah, 17, a laborer; Lydia Winehold, 17; Amos Gockley, 17, a shoemaker; and property valued at $3700.*

### John Bixler

- 71 acres improved and 15 acres unimproved land valued at $1700.
- Farm machinery valued at $158.
- 2 horses, 3 milch cows, 4 swine, valued at $157.
- 65 bu. wheat, 7 bu. rye, 60 bu. Indian corn, 375 bu. oats, 30 bu. Irish potatoes, 2 bu. buckwheat, 300 lbs. butter, 15 tons hay, 1 bu. clover seed, 10 lbs. flax, 1 bu. flaxseed.
- Animals slaughtered valued at $40.

*Federal census lists John, 29; Levina, 26; Elizabeth, 4; Samuel, 3; John, 1; and property valued at $1700.*

### John McColm

- 46 acres improved and 6 acres unimproved land valued at $2000.
- Farm machinery valued at $50.
- 2 horses, 4 milch cows, 5 other cattle, 3 swine, valued at $200.
- 90 bu. wheat, 10 bu. rye, 100 bu. Indian corn, 150 bu. oats, 23 bu. Irish potatoes, 6 bu. buckwheat, 100 lbs. butter, 8 tons hay, 1 bu. clover seed, 1 bu. other grass seed.
- Homemade manufactures valued at $6.
- Animals slaughtered valued at $35.

*Federal census lists John, 50; Mary, 54; David, 29, a carpenter; Mary, 25; John, 15, a laborer; William, 9; John Oberholtzer, 2. No property value given.*

### Peter Fankhauser

- 78 acres improved and 30 acres unimproved land valued at $2700.
- Farm machinery valued at $80.
- 2 horses, 4 milch cows, 6 swine, valued at $180.
- 30 bu. wheat, 9 bu. rye, 50 bu. Indian corn, 15 bu. oats, 18 bu. Irish potatoes, 6 bu. buckwheat, 100 lbs. butter, 9 tons hay.
- Animals slaughtered valued at $30.

*Federal census lists Peter, 72; Magdalena, 67; Caroline Fry, 15; and property valued at $2700.*

### David Bixler

- 50 acres improved and 25 acres unimproved land valued at $1400.
- Farm machinery valued at $28.
- 2 horses, 2 milch cows, 2 other cattle, 3 swine, valued at $136.
- 13 bu. wheat, 22 bu. rye, 45 bu. Indian corn, 50 bu. oats, 12 lbs. wool*, 8 bu. Irish potatoes, 5 bu. buckwheat, 80 lbs. butter, 7 tons hay, 7 lbs. flax, 1 bu. flaxseed.
- Animals slaughtered valued at $29.

*Note: Could possibly belong to the following entry, Samuel Fankhauser).

*Federal census lists David, 42; Mary, 49; Suzanna, 19; Mary Canny, 5; and property valued at $1400.*

### Samuel Fankhauser

- 85 acres improved and 20 acres unimproved land valued at $1800.
- Farm machinery valued at $30.
- 3 horses, 3 milch cows, 2 other cattle, 2 sheep, 4 swine, valued at $121.
- 35 bu. wheat, 25 bu. rye, 50 bu. Indian corn, 50 bu. oats, 15 bu. Irish potatoes, 50 lbs. butter, 6 tons hay, 6 lbs. flax, 1 bu. flaxseed.
- Animals slaughtered valued at $40.

*Federal census lists Samuel, 37; Elizabeth, 35; Chambers, 8 (male); Albert, 6; Catherine, 3; and property valued at $1800.*

### Isaac Messner

- 32 acres improved and 8 acres unimproved land valued at $4000.
- Farm machinery valued at $100.
- 4 horses, 4 milch cows, 4 swine, valued at $216.
- 50 bu. wheat, 20 bu. Indian corn, 80 bu. oats, 25 bu. Irish potatoes, 100 lbs. butter, 18 tons hay.
- Homemade manufactures valued at $3.
- Animals slaughtered valued at $45.

*Federal census lists Isaac, 42, innkeeper; Harriet, 33; Alexander, 16; Catherine, 14; Lydia, 12; Mary Ann, 7; Elizabeth, 6; Isaac, 3; Lehman, 11/12; and property valued at $4000.*

**Jacob Witman**
- 86 acres improved and 15 acres unimproved land valued at $2000.
- Farm machinery valued at $175.
- 2 horses, 5 milch cows, 4 swine, valued at $130.
- 176 bu. wheat, 47 bu. rye, 300 bu. Indian corn, 500 bu. oats, 45 bu. Irish potatoes, 9 bu. buckwheat, 200 lbs. butter, 12 tons hay, 2 bu. clover seed, 1 bu. other grass seed, 15 lbs. flax, 2 bu. flaxseed.
- Animals slaughtered valued at $200.

*Federal census lists Jacob, 32; Ann, 29; Henry, 9; Lydia, 6; Fay Anne, 4; Catherine Ann, 1; and property valued at $2000.*

**William Coldren**
- 92 acres improved and 8 acres unimproved land valued at $2000.
- Farm machinery valued at $150.
- 6 horses, 5 milch cows, 4 other cattle, 4 sheep, 2 swine, valued at $320.
- 20 bu. wheat, 31 bu. rye, 100 bu. Indian corn, 50 bu. oats, 19 lbs. wool, 25 bu. Irish potatoes, 200 lbs. butter, 8 tons hay, 3 bu. clover seed.
- Animals slaughtered valued at $59.

*Federal census lists William, 66, shoemaker; Magdalene, 67; Lydia, 32; Esther Fankhauser, 72; Adam Coldren, 10; Mary, 1; and property valued at $2000. (This may be the census record for William Coldren, below.)*

**William Coldren**
- 75 acres improved and 25 acres unimproved land valued at $2000.
- 1 horse, 2 milch cows, 5 swine, valued at $50.
- 75 bu. wheat, 25 bu. rye, 100 bu. Indian corn, 150 bu. oats, 20 bu. Irish potatoes, 100 lbs. butter, 12 tons hay, 1 bu. clover seed.
- Animals slaughtered valued at $25.

*Federal census lists William, 28; Lydia, 27; Abriam, 3; Owen, 1; William Coldren, 14; Sarah Lutz, 13; and property valued at $2000. (This may be the census record for William Coldren, above.)*

**Jacob Coldren**
- 50 acres improved and 20 acres unimproved land valued at $2000.
- Farm machinery valued at $50.
- 3 horses, 4 milch cows, 4 other cattle, 3 sheep, 3 swine, valued at $220.
- 50 bu. wheat, 60 bu. rye, 50 bu. Indian corn, 100 bu. oats, 10 lbs. wool, 2 bu. peas and beans, 50 bu. Irish potatoes, 250 lbs. butter, 15 tons hay, 2 bu. clover seed, 1 bu. other grass seed, 18 lbs. flax, 4 bu. flaxseed.
- Homemade manufactures valued at $4.
- Animals slaughtered valued at $65.

*Federal census lists Jacob, 49; Elizabeth, 47; Fayanne, 17; Henry, 16; Elias, 15; Jacob, 12; Levi, 8; Edward, 5; and property valued at $2000.*

**Isaac Bowman**
- 44 acres improved and 3 acres unimproved land valued at $1800.
- Farm machinery valued at $65.
- 2 horses, 3 milch cows, 3 other cattle, 4 swine, valued at $190.
- 35 bu. wheat, 40 bu. rye, 100 bu. Indian corn, 175 bu. oats, 100 bu. Irish potatoes, 2 bu. sweet potatoes, 350 lbs. butter, 12 tons hay, 2 bu. clover seed.
- Animals slaughtered valued at $65.

*Federal census lists Isaac, 32; Leah, 35; Elias, 9; Samuel, 8; Nancy, 5; Suzanne, 3; Sarah Winehold, 13; and property valued at $1800.*

**Jonas Good**
- 78 acres improved and 10 acres unimproved land valued at $2500.
- Farm machinery valued at $100.
- 5 horses, 5 milch cows, 5 other cattle, 5 swine, valued at $545.
- 76 bu. wheat, 60 bu. rye, 120 bu. Indian corn, 280 bu. oats, 60 bu. Irish potatoes, 400 lbs. butter, 20 tons hay, 1 bu. clover seed, 6 bu. other grass seed, 10 lbs. flax, 1 bu. flaxseed.
- Homemade manufactures valued at $4.
- Animals slaughtered valued at $88.

*Federal census lists Jonas, 46; Esther, 44; Catherine, 16; Amos, 14; Elizabeth, 11; Levina, 8; William, 4; and property valued at $2500. (This may be the census record for Jonas Good, below.)*

**Jonas Good**
- 60 acres improved and 17 acres unimproved land valued at $2500.
- Farm machinery valued at $300.
- 4 horses, 5 milch cows, 6 other cattle, 3 swine, valued at $346.

- 50 bu. wheat, 110 bu. rye, 100 bu. Indian corn, 125 bu. oats, 80 bu. Irish potatoes, 300 lbs. butter, 16 lbs. cheese, 25 tons hay, 4 bu. clover seed, 3 bu. other grass seed, 8 lbs. flax, 2 bu. flaxseed.
- Homemade manufactures valued at $5.
- Animals slaughtered valued at $70.

*Federal census lists Jonas, 37; Catherine, 35; Mary, 32; John Litterling, 16; and property valued at $2500. (This may be the census record for Jonas Good, above.)*

### Benjamin Musser
- 36 acres improved and 11 acres unimproved land valued at $1500.
- Farm machinery valued at $75.
- 1 horse, 4 milch cows, 3 other cattle, 3 swine, valued at $100.
- 40 bu wheat, 25 bu. rye, 75 bu. Indian corn, 50 bu. oats, 1 bu. peas and beans, 100 bu. Irish potatoes, 1 bu. sweet potatoes, 300 lbs. butter, 8 tons hay, 3 bu. clover seed.
- Animals slaughtered valued at $45.

*Federal census lists Benjamin, 43; Elizabeth, 36; Sarah, 12; Lydia, 8; Elizabeth, 3; Samuel, 1; and property valued at $1500.*

### Elias Good
- 40 acres improved and 10 acres unimproved land valued at $1500.
- Farm machinery valued at $85.
- 2 horses, 5 milch cows, 3 other cattle, 4 swine, valued at $250.
- 50 bu. wheat, 18 bu. rye, 60 bu. Indian corn, 70 bu. oats, 30 bu. Irish potatoes, 300 lbs. butter, 12 lbs. cheese, 15 tons hay, 1 bu. other grass seed.
- Animals slaughtered valued at $60.

*Federal census lists Elias, 47; Fanny, 43; Elizabeth Hiepner, 16; Mary Schwerd, 4; and property valued at $1500.*

### John Steffy
- 35 acres improved and 12 acres unimproved land valued at $2000.
- Farm machinery valued at $75.
- 2 horses, 4 milch cows, 4 other cattle, 4 swine, valued at $205.
- 20 bu. wheat, 50 bu. rye, 60 bu. Indian corn, 70 bu. oats, 20 bu. Irish potatoes, 200 lbs. butter, 12 tons hay, 2 bu. other grass seed, 30 lbs. flax, 1 bu. flaxseed.
- Homemade manufactures valued at $9.
- Animals slaughtered valued at $60.

*Federal census lists John, 62; Eve, 62; Lydia, 28; Nancy, 26; and property valued at $2000.*

### Peter Steffy
- 28 acres improved and 3 acres unimproved land valued at $1000.
- Farm machinery valued at $70.
- 1 horse, 3 milch cows, 1 other cattle, 5 sheep, 4 swine, valued at $300.
- 25 bu. wheat, 25 bu. rye, 30 bu. Indian corn, 8 lbs. wool, 30 bu. Irish potatoes, 100 lbs. butter, 5 tons hay, 9 lbs. flax, 1 bu. flaxseed.
- Homemade manufactures valued at $3.
- Animals slaughtered valued at $20.

*Federal census lists Peter, 52; Margaret, 39; Manassah, 11; Caroline, 8; Amanda, 5; and property valued at $1000.*

### Daniel Overholtzer
- 110 acres improved and 41 acres unimproved land valued at $4000.
- Farm machinery valued at $200.
- 5 horses, 5 milch cows, 8 other cattle, 8 swine, valued at $492.
- 150 bu. wheat, 30 bu. rye, 50 bu. Indian corn, 150 bu. oats, 60 bu. Irish potatoes, 1 bu. sweet potatoes, 500 lbs. butter, 40 tons hay.
- Animals slaughtered valued at $140.

*Federal census lists Daniel, 42; Sarah, 37; Suzannah, 9; John, 2; Abraham O., 39; Esther, 44; and property valued at $4000.*

### Henry Gring
- 18 acres improved and 18 acres unimproved land valued at $600.
- Farm machinery valued at $50.
- 1 horse, 3 milch cows, 1 other cattle, 3 sheep, 3 swine, valued at $74.
- 25 bu. wheat, 45 bu. oats, 27 lbs. wool, 70 bu. Irish potatoes, 12 bu. buckwheat, $10 value market garden's produce, 150 lbs. butter, 5 tons hay, 12 lbs. flax, 1 bu. flaxseed.
- Homemade manufactures valued at $10.
- Animals slaughtered valued at $37.

*Federal census lists Henry, 39; Elizabeth, 38; Susanna, 11; Allison, 8 (male); John, 6; and property valued at $600.*

## Benjamin Kessler
- 65 acres improved and 15 acres unimproved land valued at $2000.
- Farm machinery valued at $50.
- 3 horses, 4 milch cows, 5 other cattle, 4 sheep, 5 swine, valued at $240.
- 20 bu. wheat, 40 bu. rye, 50 bu. Indian corn, 60 bu. oats, 14 lbs. wool, 30 bu. Irish potatoes, 400 lbs. butter, 12 tons hay, 2 bu. other grass seed, 12 lbs. flax, 1 bu. flaxseed.
- Homemade manufactures valued at $9.
- Animals slaughtered valued at $71.

*Federal census lists Benjamin, 48; Susanna, 43; Elizabeth, 16; Mary, 9; Hiram, 5; and property valued at $2000.*

## David Roth
- 54 acres improved and 31 acres unimproved land valued at $2000.
- Farm machinery valued at $140.
- 4 horses, 4 milch cows, 5 other cattle, 4 swine, valued at $380.
- 40 bu. wheat, 40 bu. rye, 100 bu. Indian corn, 150 bu. oats, 40 bu. Irish potatoes, 300 lbs. butter, 30 tons hay, 2 bu. clover seed, 26 lbs. flax, 2 bu. flaxseed.
- Homemade manufactures valued at $5.
- Animals slaughtered valued at $52.

*Federal census lists David, 63; Elizabeth, 58; Isaac, 33, a farmer; Franklin, 6; and property valued at $2000.*

## William von Nieda
- 40 acres improved and 25 acres unimproved land valued at $6000.
- Farm machinery valued at $150.
- 3 horses, 3 milch cows, 1 other cattle, 5 swine, valued at $332.
- 66 bu. wheat, 11 bu. rye, 40 bu. Indian corn, 68 bu. oats, 20 bu. Irish potatoes, 200 lbs. butter, 15 tons hay.
- Homemade manufactures valued at $8.
- Animals slaughtered valued at $60.

*Federal census lists William, 29, a miller; Eliza, 27; Monroe, 2; Martha, 4; Ann, 1/12; Elizabeth Coldren, 14; Susanna von Nieda, 64; and property valued at $12,000.*

## Christian Wenger
- 84 acres improved and 19 acres unimproved land valued at $5000.
- Farm machinery valued at $150.
- 2 horses, 6 milch cows, 4 working cows, 5 other cattle, 7 swine, valued at $229.
- 250 bu. wheat, 10 bu. rye, 150 bu. Indian corn, 250 bu. oats, 20 lbs. wool, 40 bu. Irish potatoes, 4 bu. sweet potatoes, 60 bu. barley, 300 lbs. butter, 30 tons hay, 2 bu. other grass seed.
- Homemade manufactures valued at $12.
- Animals slaughtered valued at $111.

*Federal census lists Christian, 47; Elizabeth, 36; Joseph, 19, a laborer; Anna, 17; Susanna, 15; Mary, 12; Peter, 10; Jonas, 9; Magdalene, 7; Isaac, 5; Christian, 4; Jacob, 1; and property valued at $5000.*

## Jacob Bicher
- 28 acres improved and 3 acres unimproved land valued at $1200.
- Farm machinery valued at $44.
- 1 horse, 2 milch cows, 3 other cattle, 1 sheep, 4 swine, valued at $86.
- 40 bu. wheat, 40 bu. rye, 25 bu. Indian corn, 50 bu. oats, 60 bu. Irish potatoes, 200 lbs. butter, 7 tons hay.
- Animals slaughtered valued at $51.

*Federal census lists Jacob, 31; Sarah, 26; Jeremiah, 5; Rebecca, 3; and property valued at $1200.*

## Samuel Lessly
- 20 acres improved land valued at $1500.
- 2 milch cows, 2 other cattle, 2 swine, valued at $60.
- 28 bu. wheat, 15 bu. rye, 90 bu. Indian corn, 115 bu. oats, 20 bu. Irish potatoes, 150 lbs. butter, 5 tons hay.
- Animals slaughtered valued at $50.

*Federal census lists Samuel, 41, a blacksmith; Margaret, 47; Rebecca, 10; Elizabeth, 8; Solomon, 2; John, 12; and property valued at $1500.*

## Daniel Gehman
- 83 acres improved and 9 acres unimproved land worth $2950.
- Farm machinery valued at $80.
- 2 horses, 4 milch cows, 1 other cattle, 4 swine, valued at $146.
- 40 bu. wheat, 75 bu. rye, 150 bu. Indian corn, 150 bu. oats, 50 bu. Irish potatoes, 250 lbs. butter, 20 tons hay, 3 bu. clover seed.
- Animals slaughtered valued at $80.

*Federal census lists Daniel, 39; Judith, 31; John Kurtz, 8.*

### Jacob Musser
- 66 acres improved and 4 acres unimproved land valued at $3000.
- Farm machinery valued at $60.
- 2 horses, 5 milch cows, 4 other cattle, 4 swine, valued at $$227.
- 50 bu. wheat, 25 bu. rye, 130 bu. Indian corn, 125 bu. oats, 100 bu. Irish potatoes, 400 lbs. butter, 14 tons hay, 1 bu. clover seed.
- Animals slaughtered valued at $71.

*Federal census lists Jacob, 28; Nancy, 30; Joseph, 7; Elizabeth, 6; Susanna, 2; and property valued at $3000.*

### Amos Schneder
- 24 acres improved and 4 acres unimproved land valued at $1000.
- Farm machinery valued at $45.
- 2 horses, 2 milch cows, 3 other cattle, 3 swine, valued at $209.
- 20 bu. wheat, 20 bu. rye, 100 bu. Indian corn, 75 bu. oats, 20 lbs. wool, 30 bu. Irish potatoes, 100 lbs. butter, 15 tons hay, 1 bu. clover seed, 25 lbs. flax, 2 bu. flaxseed.
- Homemade manufactures valued at $7.
- Animals slaughtered valued at $42.

*Federal census lists Amos, 42; Catherine, 42; Baltzer, 18, a laborer; Rebecca, 21; Elizabeth, 14; Mary, 11; Catherine, 3/12; and property valued at $1000.*

### Christian Pleam
- 96 acres improved and 17 acres unimproved land valued at $3000.
- Farm machinery valued at $75.
- 3 horses, 4 milch cows, 5 other cattle, 5 swine, valued at $280.
- 80 bu. wheat, 50 bu. rye, 200 bu. Indian corn, 225 bu. oats, 20 bu. Irish potatoes, 400 lbs. butter, 23 tons hay, 3 bu. other grass seed, 10 lbs. flax, 1 bu. flaxseed.
- Homemade manufactures valued at $8.
- Animals slaughtered valued at $80.

*Federal census lists Christian Ream, 52; Elizabeth Ream, 45; Barbara Ream, 42; Israel Messner, 18, a laborer; and property valued at $3000.*

### Moses Messner
- 54 acres improved and 10 acres unimproved land valued at $1300.
- Farm machinery valued at $50.
- 2 horses, 4 milch cows, 1 other cattle, 4 swine, valued at $180.
- 70 bu. rye, 80 bu. Indian corn, 20 bu. oats, 20 bu. Irish potatoes, 100 lbs. butter, 8 tons hay, 12 lbs. flax, 1 bu. flaxseed.
- Animals slaughtered valued at $33.

*Federal census lists Moses, 25; Elizabeth, 23; Elizabeth Foltz, 58; and property valued at $1300.*

### Daniel Messner
- 55 acres improved and 25 acres unimproved land valued at $1600.
- Farm machinery valued at $135.
- 2 horses, 4 milch cows, 3 other cattle, 4 swine, valued at $219.
- 50 bu. wheat, 30 bu. rye, 80 bu. Indian corn, 80 bu. oats, 50 bu. Irish potatoes, 150 lbs. butter, 15 tons hay, 1 bu. clover seed, 2 bu. other grass seed, 18 lbs. flax, 6 bu. flaxseed.
- Homemade manufactures valued at $5.
- Animals slaughtered valued at $62.

*Federal census lists Daniel, 54; Catherine, 48; Elizabeth, 28; Martin, 17, a laborer; Magdaline, 13; and property valued at $1600.*

### Samuel Messner
- 72 acres improved and 18 acres unimproved land valued at $4000.
- Farm machinery valued at $300.
- 4 horses, 5 milch cows, 1 other cattle, 6 swine, valued at $408.
- 150 bu. wheat, 30 bu. rye, 250 bu. Indian corn, 275 bu. oats, 30 bu. Irish potatoes, 200 lbs. butter, 20 tons hay, 15 lbs. flax, 1 bu. flaxseed.
- Homemade manufactures valued at $5.
- Animals slaughtered valued at $67.

*Federal census lists Samuel, 39; Lydia, 43; Samuel, 2, Isaac Coldren, 21, a laborer; Lydia Coldren, 23; Mary Messner, 65; and property valued at $4000.*

### Joseph Kooser
- 28 acres improved and 2 acres unimproved land valued at $1800.
- Farm machinery valued at $100.

- 1 horse, 3 milch cows, 3 swine, valued at $175.
- 40 bu. wheat, 40 bu. rye, 100 bu. Indian corn, 160 bu. oats, 20 bu. Irish potatoes, 100 lbs. butter, 6 tons hay, 15 lbs. flax, 1 bu. flaxseed.
- Homemade manufactures valued at $4.
- Animals slaughtered valued at $60.

*Federal census lists Joseph, 34; Mary, 31; Lehman, 1; and property valued at $1800.*

## George Martin

- 32 acres improved and 12 acres unimproved land valued at $2500.
- Farm machinery valued at $100.
- 2 horses, 3 milch cows, 6 swine, valued at $170.
- 30 bu. wheat, 12 bu. rye, 25 bu. Indian corn, 110 bu. oats, 18 bu. Irish potatoes, 400 lbs. butter, 2 tons hay.
- Animals slaughtered valued at $67.

*Federal census lists George, 35; Levina, 26; Anna, 8; Mary, 7; Moses, 5; George, 3; Joseph, 2; Matilda, 6/12; John Marn(?), 30, a laborer; Sarah Laush, 16; and property valued at $2500.*

## Abraham Weaver

- 62 acres improved and 10 acres unimproved land valued at $3000.
- Farm machinery valued at $93.
- 4 horses, 4 milch cows, 7 other cattle, 6 swine, valued at $272.
- 110 bu. wheat, 40 bu. rye, 90 bu. Indian corn, 220 bu. oats, 10 bu. Irish potatoes, 2 bu. barley, 180 lbs. butter, 10 tons hay, 9 bu. clover seed.
- Animals slaughtered valued at $95.

*Federal census lists Abraham, 67; Mary, 64; Abraham, 25, a laborer; Mary, 22; Christian Fankhauser, 11; and property valued at $3000.*

## John Lessly

- 36 acres improved and 7 acres unimproved land valued at $2000.
- Farm machinery valued at $60.
- 2 horses, 2 milch cows, 5 other cattle, 3 swine, valued at $120.
- 80 bu. wheat, 25 bu. rye, 75 bu. Indian corn, 100 bu. oats, 20 bu. Irish potatoes, 100 lbs. butter, 5 tons hay.
- Animals slaughtered valued at $60.

*Federal census lists John, 53; Susanna, 39; Elizabeth, 16; Lydia, 12; William, 7; and property valued at $2000.*

## John Miller

- 68 acres improved and 30 acres unimproved land valued at $1800.
- Farm machinery valued at $25.
- 2 horses, 4 milch cows, 6 other cattle, 5 sheep, 2 swine, valued at $80.
- 33 bu. wheat, 13 bu. rye, 120 bu. Indian corn, 150 bu. oats, 11 lbs. wool, 10 bu. Irish potatoes, 100 lbs. butter, 3 tons hay.
- Animals slaughtered valued at $20.

*Federal census lists John, 38; Sarah, 33; Christian, 13; Catherine, 10; John, 8; Rachel, 5; Susanna, 4; Cyrus, 2; Bair, 2/12; and property valued at $1800.*

## Elizabeth Haldeman

- 32 acres improved and 5 acres unimproved land valued at $1800.
- Farm machinery valued at $55.
- 1 horse, 2 milch cows, 1 other cattle, 3 swine, valued at $144.
- 30 bu. wheat, 20 bu. rye, 80 bu. Indian corn, 80 bu. oats, 25 bu. Irish potatoes, 150 lbs. butter, 4 tons hay.
- Animals slaughtered valued at $37.

*Federal census lists Elizabeth, 45; Elias, 23, a laborer; Elizabeth, 18; and property valued at $1800.*

## Jacob Renninger

- 70 acres improved and 30 acres unimproved land valued at $3500.
- Farm machinery valued at $100.
- 3 horses, 6 milch cows, 7 other cattle, 8 swine, valued at $140.
- 100 bu. wheat, 25 bu. rye, 100 bu. Indian corn, 150 bu. oats, 40 bu. Irish potatoes, 300 lbs. butter, 30 tons hay, 2 bu. clover seed, 2 bu. other grass seed, 22 lbs. flax, 2 bu. flaxseed.
- Homemade manufactures valued at $6.
- Animals slaughtered valued at $100.

*Federal census lists Jacob, 69; Catherine, 65; and property valued at $3500.*

### Jonas Good
• 65 acres improved and 35 acres unimproved land valued at $2800.
• Farm machinery valued at $100.
• 2 horses, 5 milch cows, 9 other cattle, 4 swine, valued at $295.
• 50 bu. wheat, 80 bu. rye, 100 bu. Indian corn, 250 bu. oats, 40 bu. Irish potatoes, 300 lbs. butter, 15 tons hay, 4 bu. clover seed, 1 bu. other grass seed.
• Animals slaughtered valued at $90.

*Federal census lists Jonas, 55; Jacob, 23, a shoemaker; Peter, 25, a drover; Nancy, 21; Solomon, 19, a laborer; Joseph, 15, a laborer; Judith, 8; and property valued at $2800.*

### Joseph Good
• 17 acres improved and 3 acres unimproved land valued at $800.
• Farm machinery valued at $45.
• 1 horse, 3 milch cows, 1 other cattle, 2 swine, valued at $90.
• 20 bu. wheat, 2 bu. rye, 30 bu. Indian corn, 30 bu. oats, 1 bu. peas and beans, 30 bu. Irish potatoes, market gardens produce valued at $2, 100 lbs. butter, 5 tons hay, 20 lbs. flax, 1 bu. flaxseed.
• Animals slaughtered valued at $49.

*Federal census lists Joseph, 55; Barbara, 49; Samuel, 21, a carpenter; Magdalena, 19; Margaret, 9; Joseph, 5; and property valued at $800.*

### Isaac Good
• 62 acres improved and 8 acres unimproved land valued at $1500.
• Farm machinery valued at $50.
• 3 horses, 5 milch cows, 5 other cattle, 4 swine, valued at $110.
• 60 bu. wheat, 30 bu. rye, 60 bu. Indian corn, 80 bu. oats, 30 bu. Irish potatoes, 2 bu. sweet potatoes, 300 lbs. butter, 8 tons hay, 15 lbs. flax, 2 bu. flaxseed.
• Animals slaughtered valued at $50.

*Federal census lists Isaac, 30; Mary, 34; Lydia, 29; Magdalena, 28; Magdalena Good, 65, a widow; Isaac Burkhart, 11; Sarah Steffy, 5; and property valued at $1500.*

### Henry Steffy
• 60 acres improved and 40 ares unimproved land valued at $3000.
• Farm machinery valued at $75.
• 4 horses, 4 milch cows, 8 other cattle, 15 sheep, 4 swine, valued at $339.
• 50 bu. wheat, 70 bu. rye, 130 bu. Indian corn, 200 bu. oats, 30 lbs. wool, 50 bu. Irish potatoes, 150 lbs. butter, 15 tons hay, 1 bu. clover seed.
• Animals slaughtered valued at $76.

*Federal census lists Henry, 58; Catherine, 54; John, 25, a laborer; Henry, 23, a laborer; Lucy, 19; Catherine, 18; William, 16, a laborer; Adam, 14; Margaret, 12; Joseph, 10; and property valued at $3000.*

### Samuel Geigly
• 45 acres improved and 5 acres unimproved land valued at $900.
• Farm machinery valued at $60.
• 2 horses, 3 milch cows, 3 other cattle, 4 swine, valued at $200.
• 26 bu. wheat, 13 bu. rye, 30 bu. Indian corn, 45 bu. oats, 50 bu. Irish potatoes, 200 lbs. butter, 10 tons hay, 12 lbs. flax, 2 bu. flaxseed.
• Animals slaughtered valued at $40.

*Federal census lists Samuel, 57; Sophia, 46; William, 22, a laborer; Eliza, 12; Samuel, 6; and property valued at $900.*

### John Shup
• 34 acres improved and 10 acres unimproved land valued at $1400.
• Farm machinery valued at $107.
• 2 horses, 5 milch cows, 6 other cattle, 7 sheep, 4 swine, valued at $162.
• 20 bu. wheat, 25 bu. rye, 20 bu. Indian corn, 25 bu. oats, 13 lbs. wool, 25 bu. Irish potatoes, 4 bu. buckwheat, 150 lbs. butter, 8 tons hay, 10 lbs. flax, 1 bu. flaxseed.
• Animals slaughtered valued at $45.

*Federal census lists John, 58; Lydia, 41; Elizabeth, 9; John, 5; and property valued at $1400.*

### Samuel Pannebecker
• 45 acres improved and 12 acres unimproved land valued at $2000.
• Farm machinery valued at $75.
• 2 horses, 4 milch cows, 1 sheep, 3 swine, valued at $200.
• 20 bu. wheat, 100 bu. Indian corn, 75 bu. oats, 20 lbs. wool, 60 bu. Irish potatoes, 100 lbs. butter, 10 tons hay.
• Animals slaughtered valued at $45.

*Federal census lists Samuel, 55; Benneville, 21, male laborer; Elizabeth Stover, 44; and property valued at $2000.*

## Godfrey Kiefer
- 80 acres improved and 17 acres unimproved land valued at $2000.
- Farm machinery valued at $25.
- 3 horses, 5 milch cows, 4 other cattle, 5 swine, valued at $130.
- 25 bu. wheat, 20 bu. rye, 50 bu. Indian corn, 45 bu. oats, 15 bu. Irish potatoes, 100 lbs. butter, 10 tons hay.
- Animals slaughtered valued at $60.

*Federal census lists Godfrey, 60; Isaac, 48, a shoemaker; Benjamin, 45, a farmer; Samuel, 44, a mason; Hannah Winehold, 36; and property valued at $2000.*

## John Stover
- 69 acres improved and 11 acres unimproved land valued at $1800.
- Farm machinery valued at $50.
- 2 horses, 3 milch cows, 7 other cattle, 3 swine, valued at $150.
- 25 bu. wheat, 25 bu. rye, 30 bu. Indian corn, 60 bu. oats, 25 bu. Irish potatoes, 150 lbs. butter, 12 tons hay.
- Animals slaughtered valued at $60.

*Federal census lists John, 50; and property valued at $1800.*

## Isaac Kessler
- 52 acres improved and 20 acres unimproved land valued at $2000.
- Farm machinery valued at $200.
- 3 horses, 4 milch cows, 2 other cattle, 3 swine, valued at $187.
- 15 bu. wheat, 30 bu. rye, 50 bu. Indian corn, 40 bu. oats, 10 lbs. wool, 25 bu. Irish potatoes, 200 lbs. butter, 10 tons hay, 1 bu. clover seed, 2 bu. other grass seed.
- Animals slaughtered valued at $45.

*Federal census lists Isaac, 37; Catherine, 37; Margaret, 6; Daniel, 2; John Musser, 12; and property valued at $2000.*

## Peter Musser
- 73 acres improved and 18 acres unimproved land valued at $2500.
- Farm machinery valued at $100.
- 3 horses, 6 milch cows, 5 other cattle, 6 swine, valued at $290.
- 90 bu. wheat, 50 bu. rye, 100 bu. Indian corn, 250 bu. oats, 50 bu. Irish potatoes, 250 lbs. butter, 15 tons hay, 2 bu. clover seed, 70 lbs. flax, 3 bu. flaxseed.
- Animals slaughtered valued at $75.

*Federal census lists Peter, 29; Sarah, 29; Franklin, 1; Barbara Hurst, 12; and property valued at $1200.*

## Peter Good
- 70 acres improved and 22 acres unimproved land valued at $2500.
- Farm machinery valued at $80.
- 4 horses, 4 milch cows, 3 other cattle, 4 sheep, valued at $260.
- 35 bu. wheat, 40 bu. rye, 50 bu. Indian corn, 50 bu. oats, 14 lbs. wool, 40 bu. Irish potatoes, 1 bu. sweet potatoes, 300 lbs. butter, 18 lbs. cheese, 7 tons hay, 2 bu. clover seed, 6 lbs. flax, 1 bu. flaxseed.
- Homemade manufactures valued at $6.
- Animals slaughtered valued at $53.

*Federal census lists Peter, 43; Elizabeth, 41; Isaac, 18, a laborer; Peter, 7; and property valued at $2500. (This census record may be for the Peter Good, below.)*

## Peter Good
- 56 acres improved and 20 acres unimproved land valued at $2500.
- Farm machinery valued at $75.
- 2 horses, 4 milch cows, 2 other cattle, 4 swine, valued at $212.
- 70 bu. wheat, 18 bu. rye, 100 bu. Indian corn, 200 bu. oats, 30 bu. Irish potatoes, 300 lbs. butter, 8 tons hay.
- Animals slaughtered valued at $70.

*Federal census lists Peter, 65; Elizabeth, 63; David, 19, a laborer; and property valued at $2500. (This census record may be for the Peter Good, above.)*

## Samuel Good
- 70 acres improved and 40 acres unimproved land valued at $2750.
- Farm machinery valued at $60.
- 2 horses, 2 milch cows, 3 swine, valued at $40.
- 80 bu. wheat, 20 bu. rye, 12 bu. Indian corn, 18 bu. oats, 8 bu. Irish potatoes, 200 lbs. butter, 5 tons hay.
- Animals slaughtered valued at $30.

*Federal census lists Samuel, 59; and property valued at $2750.*

### John Becker
• 67 acres improved and 6 acres unimproved land valued at $2000.
• Farm machinery valued at $65.
• 2 horses, 4 milch cows, 5 other cattle, and 4 swine, valued at $194.
• 25 bu. wheat, 60 bu. rye, 100 bu. Indian corn, 300 bu. oats, 50 bu. Irish potatoes, 200 lbs. butter, 10 lbs. cheese, 12 tons hay, 1 bu. other grass seed.
• Animals slaughtered valued at $73.
*Federal census lists John, 50; Barbara, 41; Peter, 16; Mary, 7; Sarah Showalter, 17; and property valued at $2000.*

### Jacob Renninger
• 19 acres improved land valued at $800.
• Farm machinery valued at $50.
• 1 horse, 2 milch cows, 3 swine, valued at $92.
• 22 bu. wheat, 4 bu. rye, 40 bu. Indian corn, 140 bu. oats, 10 bu. Irish potatoes, 200 lbs. butter, 4 tons hay, 18 lbs. flax, 1 bu. flaxseed.
• Homemade manufactures valued at $4.
• Animals slaughtered valued at $35.
*Federal census lists Jacob, 39, a blacksmith; Catherine, 34; Caroline, 9; John Jacob, 7; Edward Davies, 2; and property valued at $800.*

### Jonas Bowman
• 47 acres improved and 10 acres unimproved land valued at $2500.
• Farm machinery valued at $65.
• 2 horses, 3 milch cows, 2 other cattle, 4 swine, valued at $240.
• 40 bu. wheat, 18 bu, rye, 40 bu. Indian corn, 140 bu. oats, 30 bu. Irish potatoes, 200 lbs. butter, 12 tons hay.
• Animals slaughtered valued at $56.
*Federal census lists Jonas, 33; Lydia, 33; John, 8; Joseph, 5; John Hutchinson, 14; Benjamin Laush, 24, a laborer; and property valued at $2500.*

### David Hutchinson
• 60 acres improved and 30 acres unimproved land valued at $3000.
• Farm machinery valued at $50.
• 2 horses, 4 milch cows, 6 other cattle, 3 swine, valued at $158.
• 50 bu. wheat, 25 bu. rye, 100 bu. Indian corn, 100 bu. oats, 50 bu. Irish potatoes, 200 lbs. butter, 15 tons hay, 5 bu. clover seed, 25 lbs. flax, 2 bu. flaxseed.
• Homemade manufactures valued at $6.
• Animals slaughtered valued at $50.
*Federal census lists David, 63; Esther, 70; Fay Anne Shaefer, 20; David Hutchinson, 18, a laborer; and property valued at $3000.*

### Jacob Brossman
• 100 acres improved and 50 acres unimproved land valued at $6000.
• Farm machinery valued at $152.
• 7 horses, 7 milch cows, 5 other cattle, 5 sheep, 4 swine, valued at $517.
• 220 bu. wheat, 50 bu. rye, 350 bu. Indian corn, 650 bu. oats, 20 lbs. wool, 80 bu. Irish potatoes, 1 bu. sweet potatoes, 600 lbs. butter, 30 tons hay, 8 bu. clover seed, 2 bu. other grass seed.
• Homemade manufactures valued at $3.
• Animals slaughtered valued at $110.
*Federal census lists Jacob, 70; Mary Ann, 67; and property valued at $6000.*

### Samuel Weinhold (Wineholt)
• 60 acres improved and 10 acres unimproved land valued at $3000.
• Farm machinery valued at $100.
• 3 horses, 5 milch cows, 5 other cattle, 4 swine, valued at $270.
• 96 bu. wheat, 104 bu. rye, 100 bu. Indian corn, 200 bu. oats, 20 bu. Irish potatoes, 250 lbs. butter, 20 tons hay, 2 bu. clover seed.
• Animals slaughtered valued at $60.
*Federal census lists Samuel, 32; Hannah, 26; Daniel, 6; Samuel, 4; Mary Ann, 3; William Griffith, 19, a laborer; Henry Wark, 19, a laborer (from Germany); and property valued at $3000.*

### Adam Stover
• 30 acres improved and 8 acres unimproved land valued at $1200.
• Farm machinery valued at $30.
• 2 horses, 3 milch cows, 4 swine, valued at $90.

- 30 bu. wheat, 12 bu. rye, 30 bu. Indian corn, 30 bu. oats, 12 bu. Irish potatoes, 100 lbs. butter, 12 tons hay, 2 bu. clover seed, 20 lbs. flax, 1 bu. flaxseed.
- Homemade manufactures valued at $8.
- Animals slaughtered valued at $58.

*Federal census lists Adam, 69; Susanna, 63; Elizabeth, 29; Jesse, 27, a laborer; Adam, 23, a mason; Samuel, 21, a mason; Noe(?), 13 (male); Elizabeth Matz, 2; and property valued at $1200.*

### George Shup

- 65 acres improved and 15 acres unimproved land valued at $800.
- Farm machinery valued at $30.
- 2 horses, 4 milch cows, 2 other cattle, 4 swine, valued at $150.
- 18 bu. wheat, 10 bu. rye, 25 bu. Indian corn, 18 bu. oats, 20 bu. Irish potatoes, 75 lbs. butter, 5 tons hay, 1 bu. clover seed, 9 lbs. flax, 1 bu. flaxseed.
- Animals slaughtered valued at $45.

*Federal census lists George, 68; Mary, 68; George, 25, a cooper; George Schneider, 17, a laborer; and property valued at $800.*

### Christian Musser

- 70 acres improved and 20 acres unimproved land valued at $1500.
- Farm machinery valued at $50.
- 2 horses, 4 milch cows, 3 other cattle, 6 sheep, 5 swine, valued at $120.
- 25 bu. wheat, 19 bu. rye, 50 bu. Indian corn, 50 bu. oats, 8 lbs. wool, 15 bu. Irish potatoes, 3 bu. sweet potatoes, 100 lbs. butter, 10 tons hay.
- Homemade manufactures valued at $5.
- Animals slaughtered valued at $45.

*Federal census lists Christian, 68; Fanny, 65; John, 29, a laborer; and property valued at $1500.*

### Daniel Musser

- 90 acres improved and 28 acres unimproved land valued at $2400.
- Farm machinery valued at $100.
- 6 horses, 5 milch cows, 6 other cattle, 5 swine, valued at $163.
- 50 bu. wheat, 30 bu. rye, 100 bu. Indian corn, 200 bu. oats, 30 bu. Irish potatoes, 200 lbs. butter, 6 lbs. cheese, 20 tons hay, 3 bu. clover seed, 1 bu. other grass seed, 15 lbs. flax, 1 bu. flaxseed.
- Homemade manufactures valued at $5.
- Animals slaughtered valued at $71.

*Federal census lists Daniel, 51; Elizabeth, 51; Isaac, 22, a laborer; David, 20, a laborer; Nancy, 18; Daniel, 8; and property valued at $2400.*

### Joseph Good

- 72 acres improved and 29 acres unimproved land valued at $3500.
- Farm machinery valued at $95.
- 3 horses, 6 cows, 6 other cattle, 5 swine, valued at $310.
- 63 bu. wheat, 63 bu. rye, 200 bu. Indian corn, 200 bu. oats, 40 bu. Irish potatoes, 2 bu. sweet potatoes, 400 lbs. butter, 10 lbs. cheese, 20 tons hay, 2 bu. clover seed, 26 lbs. flax, 1 bu. flaxseed.
- Homemade manufactures valued at $7.
- Animals slaughtered valued at $72.

*Federal census lists Joseph, 72; Catherine, 34; Mary, 32; Jonathan, 15, a laborer; and property valued at $3500.*

### Samuel Laush

- 78 acres improved and 33 acres unimproved land valued at $1500.
- Farm machinery valued at $150.
- 2 horses, 3 milch cows, 2 other cattle, 4 sheep, 3 swine, valued at $150.
- 50 bu. wheat, 5 bu. rye, 50 bu. Indian corn, 100 bu. oats, 8 lbs. wool, 100 bu. Irish potatoes, value of orchard products at $1,100 lbs. butter, 6 tons hay.
- Homemade manufactures valued at $7.
- Animals slaughtered valued at $40.

*Federal census lists Samuel, 52; Nancy, 42; Louisa, 13; and property valued at $1500.*

### John Laush

- 20 acres improved and 22 acres unimproved land valued at $800.
- Farm machinery valued at $50.
- 1 horse, 3 milch cows, 4 swine, valued at $98.
- 35 bu. wheat, 5 bu. rye, 20 bu. Indian corn, 32 bu. oats, 40 bu. Irish potatoes, value of orchard products at $2, 75 lbs. butter, 10 tons hay, 2 bu. clover seed.

• Animals slaughtered valued at $30.
*Federal census lists John, 62; Magdalena, 63; Levi, 33, a farmer; Mary, 34; Lydia Ann, 8; John, 4; Mary, 3; Eliza, 7/12; and property valued at $1000.*

**Joel Brossman**
• 85 acres improved and 15 acres unimproved land valued at $3500.
• Farm machinery valued at $150.
• 4 horses, 5 milch cows, 5 other cattle, 5 swine, valued at $250.
• 100 bu. wheat, 18 bu. rye, 100 bu. Indian corn, 200 bu. oats, 50 bu. Irish potatoes, 250 lbs. butter, 30 tons hay, 5 bu. clover seed, 1 bu. other grass seed.
• Animals slaughtered valued at $60.
*Federal census lists Joel, 35; Mary, 31; Henry, 11; Caroline, 9; Barbara Ann, 1; John Ehring, 17, a laborer; and property valued at $3500.*

**Isaac Renninger**
• 22 acres improved land valued at $1500.
• Farm machinery valued at $110.
• 1 horse, 2 milch cows, 2 swine, valued at $90.
• 8 bu. wheat, 25 bu. rye, 20 bu. Indian corn, 110 bu. oats, 8 bu. Irish potatoes, 150 lbs. butter, 7 tons hay.
• Animals slaughtered valued at $35.
*Federal census lists Isaac, 35; Lydia, 23; Martin, 8; Mary, 6; Cassiah, 1; and property valued at $1300.*

**Samuel Becker**
• 42 acres improved and 8 acres unimproved land valued at $2000.
• Farm machinery valued at $50.
• 2 horses, 4 milch cows, 1 other cattle, 4 swine, valued at $160.
• 40 bu. wheat, 40 bu. rye, 90 bu. Indian corn, 200 bu. oats, 25 bu. Irish potatoes, 1 bu. sweet potatoes, 400 lbs. butter, 9 tons hay, 1 bu. clover seed.
• Animals slaughtered valued at $60.
*Federal census lists Samuel, 46; Mary, 42; Eliza, 19; Judith Good, 46; Lydia Good, 7; and property valued at $2000.*

**Peter Becker**
• 40 acres improved and 10 acres unimproved land valued at $2200.
• Farm machinery valued at $85.
• 3 horses, 3 milch cows, 3 other cattle, 3 swine, valued at $168.
• 40 bu. wheat, 30 bu. rye, 75 bu. Indian corn, 110 bu. oats, 35 bu. Irish potatoes, 2 bu. sweet potatoes, 2 bu. barley, 300 lbs. butter, 10 tons hay, 1 bu. clover seed, 1 lb. hops, 9 lbs. flax, 1 bu. flaxseed.
• Homemade manufactures valued at $3.
• Animals slaughtered valued at $80.
*Federal census lists Peter, 49; Esther, 49; Catherine, 12; and property valued at $2200.*

**Amos Hutchinson**
• 24 acres improved and 5 acres unimproved land valued at $1200.
• Farm machinery valued at $20.
• 1 horse, 3 milch cows, 2 other cattle, 8 sheep, 4 swine, valued at $69.
• 10 bu. wheat, 10 bu. rye, 20 bu. Indian corn, 25 bu. oats, 6 lbs. wool, 15 bu. Irish potatoes, 50 lbs. butter, 5 tons hay, 10 lbs. flax, 2 bu. flaxseed.
• Animals slaughtered valued at $75.
*Federal census lists Amos, 44; Nancy, 45; Peter, 13; Catherine, 7; and property valued at $1200.*

**Jacob Laush**
• 24 acres improved and 20 acres unimproved land valued at $1000.
• Farm machinery valued at $75.
• 1 horse, 3 milch cows, 3 swine, valued at $158.
• 35 bu. wheat, 5 bu. rye, 50 bu. Indian corn, 55 bu. oats, 45 bu. Irish potatoes, 200 lbs. butter, 6 tons hay.
• Animals slaughtered valued at $53.
*Federal census lists Jacob, 46, a shoemaker; Isaac, 15; Sophia, 13; Elizabeth Laush, 44; and property valued at $1000.*

# Missed on Agriculture List

### Martin Myer
*Federal census lists Martin, 25, a farmer; Esther, 19; Samuel, 23.*

### Samuel Foltz
*Federal census lists Samuel, 29, a farmer; Elizabeth, 31; Lucetta, 9; Elizabeth, 5; Abraham, 2; Peter, 5/12; Moses Huet, 9.*

### Michael Wolf
*Federal census lists Michael, 36, a farmer; Lydia, 30; Lazarus, 8; Elizabeth, 7; Michael, 5; Lydia, 2; Magdalene, 5/12.*

### Henry S. Davis
*Federal census lists Henry, 26, a farmer; Susanna, 24; Elmina Elizabeth, 2; Lydia Ann, 1; Samuel Showalter, 16, a laborer; Jesse Bixler, 9.*

### Amos Beam
*Federal census lists Amos, 26, a farmer; Mary, 26; Louisa, 2; John, 1; Elizabeth Seitzinger, 10.*

### William Renninger
*Federal census lists William, 24, a farmer; Susanna, 24; Nathan, 1; Barbara Messner, 17; Benjamin Wolf, 27, a laborer.*

### Deitrich Hartman
*Federal census lists Deitrich, 29, a farmer (from Germany); Elizabeth, 33; Martin Geigley, 6.*

### Moses Bowman
*Federal census lists Moses, 26, a farmer; Mary, 31.*

# Brecknock Township—1860

## David Hoffman
- 20 acres improved and 8 acres unimproved land valued at $1200.
- Farm machinery valued at $50.
- 1 horse, 3 milch cows, 2 other cattle, 3 swine, valued at $150.
- 47 bu. wheat, 45 bu. Indian corn, 16 bu. oats, 18 bu. Irish potatoes, orchard products valued at $6, 200 lbs. butter, 4 tons hay.
- Animals slaughtered valued at $60.

*Federal census lists David, 50, farm laborer; Catherine, 39; David, 16, a farm laborer; Salinda, 14; Henry, 12; Israel, 10; Samuel, 9; William, 7; Eliza, 4; Catherine, 1; and property valued at $1300.*

## Ruben Wolfskill
- 20 acres improved and 3 acres unimproved land valued at $1000.
- Farm machinery valued at $50.
- 2 horses, 2 milch cows, 4 swine, valued at $200.
- 45 bu. wheat, 5 bu. rye, 35 bu. Indian corn, 20 bu. oats, 12 bu. Irish potatoes, 100 lbs. butter, 4 tons hay.
- Animals slaughtered valued at $40.

*Federal census lists Ruben, 44, a day laborer; Catherine, 43; Jacob, 20, a day laborer; Ruben, 17, a day laborer; Amos, 10; and property valued at $1000.*

## Michael G. Grove
- 40 acres improved and 28 acres unimproved land valued at $3500.
- Farm machinery valued at $250.
- 2 horses, 5 asses and mules, 4 milch cows, 6 other cattle, 6 swine, valued at $680.
- 175 bu. wheat, 5 bu. rye, 450 bu. Indian corn, 200 bu. oats, 11 lbs. tobacco, 50 bu. Irish potatoes, orchard products valued at $35, 260 lbs. butter, 25 tons hay, 5 bu. clover seed, 7 bu. other grass seed, 25 lbs. honey.
- Animals slaughtered valued at $105.

*Federal census lists Michael, 23; Eliza, 24; George F., 2; Susanna Gunaway, 12; Josiah Fry, 18, farm laborer; and property valued at $3500.*

## John Hoffman
- 100 acres improved and 50 acres unimproved land valued at $3500.
- Farm machinery valued at $70.
- 5 horses, 4 milch cows, 8 other cattle, 4 swine, valued at $600.
- 300 bu. wheat, 150 bu. Indian corn, 200 bu. oats, 15 bu. Irish potatoes, orchard products valued at $5, 300 lbs. butter, 25 tons hay, 2 bu. clover seed, 5 lbs. honey.
- Animals slaughtered valued at $80.

*Federal census lists John, 64; Catherine, 65; Samuel, 26 (insane); John, 23, a farm laborer; George, 20, a farm laborer; and property valued at $4000.*

## Rachel Pheifer
- 35 acres improved and 3 acres unimproved land valued at $1000.
- Farm machinery valued at $50.
- 3 horses, 2 milch cows, 3 other cattle, 2 swine, valued at $300.
- 50 bu. wheat, 10 bu. rye, 30 bu. Indian corn, 40 bu. oats, 20 bu. Irish potatoes, 100 lbs. butter, 5 tons hay, 1 bu. clover seed.
- Animals slaughtered valued at $65.

*Federal census lists Rachel, 65, a widow; George, 26, a farm laborer; and property valued at $1000.*

## Jacob Fankhauser
- 12 acres improved land valued at $1500.
- Farm machinery valued at $50.
- 1 horse, 2 milch cows, 2 other cattle, 6 swine, valued at $150.
- 52 bu. wheat, 80 bu. Indian corn, 70 bu. oats, 15 lbs. tobacco, 25 bu. Irish potatoes, 2 bu. sweet potatoes, 7 bu. buckwheat, 200 lbs. butter, 5 tons hay.
- Animals slaughtered valued at $105.

*Federal census lists Jacob, 49, a master carpenter; Elizabeth, 53; Mary, 13; and property valued at $1500.*

### Jacob Holsinger
• 45 acres improved and 14 acres unimproved land valued at $2300.
• Farm machinery valued at $71.
• 2 horses, 2 milch cows, 5 other cattle, 9 sheep, 3 swine, valued at $160.
• 70 bu. wheat, 15 bu. rye, 150 bu. Indian corn, 100 bu. oats, 27 lbs. wool, 20 bu. Irish potatoes, 9 bu. sweet potatoes, 15 bu. buckwheat, 200 lbs. butter, 12 tons hay, 1 bu. clover seed, 1 bu. other grass seed.
• Animals slaughtered valued at $30.
*Federal census lists Jacob, 80; Catherine, 50; John, 36, a farm laborer; Mary, 34; and property valued at $2300.*

### Richard Adams
• 9 acres improved land valued at $1000.
• 2 milch cows, 2 swine, valued at $50.
• 42 bu. wheat, 75 bu. Indian corn, 20 bu. Irish potatoes, 200 lbs. butter, 2 tons hay, 1 bu. clover seed.
• Animals slaughtered valued at $45.
*Federal census lists Richard, 32, a master mason; Anna, 32; Elmina, 11; Richard, 8; John, 8; Susanna, 7; Eliza, 2; and property valued at $1000.*

### Jacob Kern
• 23 acres improved and 8 acres unimproved land valued at $1500.
• Farm machinery valued at $100.
• 2 horses, 4 milch cows, 4 other cattle, 2 swine, valued at $200.
• 30 bu. wheat, 3 bu. rye, 30 bu. Indian corn, 18 bu. Irish potatoes, 300 lbs. butter, 8 tons hay.
• Animals slaughtered valued at $55.
*Federal census lists Jacob, 49; Phianna, 45; Solomon, 19, a day laborer; Edward, 18, a day laborer; Henry, 13; Richard, 11; and property valued at $1500.*

### Curtis Fankhauser
• 17 acres improved land valued at $2000.
• Farm machinery valued at $35.
• 1 horse, 3 milch cows, 4 swine, valued at 170.
• 45 bu. wheat, 70 bu. Indian corn, 60 bu. oats, 30 bu. Irish potatoes, 100 lbs. butter, 3 tons hay.
• Animals slaughtered valued at $60.
*Federal census lists Curtis, 52, a miller; Manda, 18; Christian, 14; Catherine Fankhauser, 22; Anna, 4; and property valued at $2000.*

### Joseph Oberholser
• 17 acres improved and 3 acres unimproved land valued at $2000.
• Farm machinery valued at $50.
• 1 horse, 4 milch cows, 1 other cattle, 8 swine, valued at $200.
• 40 bu. wheat, 34 bu. Indian corn, 40 bu. oats, 35 bu. Irish potatoes, 300 lbs. butter, 7 tons hay.
• Animals slaughtered valued at $127.
*Federal census lists Joseph, 39, a master miller; Magdelina, 32; Maria, 15; Daniel, 12; Abraham, 10; Joseph, 8; Susanna, 6; Catherine, 3; Martin, 1/12; Hetty Ann Miller, 17, serving; and property valued at $9000.*

### Joseph Wenger
• 90 acres improved and 30 acres unimproved land valued at $6500.
• Farm machinery valued at $200.
• 4 horses, 7 milch cows, 13 other cattle, 8 swine, valued at $600.
• 300 bu. wheat, 6 bu. rye, 575 bu. Indian corn, 450 bu. oats, 20 bu. Irish potatoes, 1 bu. sweet potatoes, 500 lbs. butter, 25 tons hay, 7 bu. clover seed, 3 bu. other grass seed.
• Animals slaughtered valued at $90.
*Federal census lists Joseph, 34; Susanna, 33; Anna, 7; Abraham, 3; Jacob, 1; Susanna, 1/12; Phianna Waiman, 14, serving; and property valued at $14,000.*

### John Oberholser
• 36 acres improved land valued at $2000.
• Farm machinery valued at $200.
• 1 horse, 6 asses and mules, 3 milch cows, 2 other cattle, 7 swine, valued at $800.
• 121 bu. wheat, 125 bu. Indian corn, 50 bu. Irish potatoes, orchard products valued at $10, 300 lbs. butter, 10 tons hay.
• Animals slaughtered valued at $87.
*Federal census lists John, 41; Lydia, 32; Anna, 2; Davis, 5/12; George Muckel, 23, a farm laborer; Lydia Datch, 20, serving; and property valued at $6000.*

### Jacob Feather, tenant

- 70 acres improved and 35 unimproved land valued at $3000.
- Farm machinery valued at $100.
- 3 horses, 5 milch cows, 5 other cattle, 15 swine, valued at $250.
- 300 bu. wheat, 15 bu. rye, 400 bu. Indian corn, 500 bu. oats, 40 bu. Irish potatoes, orchard products valued at $10, 300 lbs. butter, 40 tons hay, 9 bu. clover seed, 2 bu. other grass seed.
- Animals slaughtered valued at $126.

*Federal census lists Jacob, 41; Catherine, 41; John, 19, a farm laborer; Mary, 17; Catherine, 12; Jacob, 8; Daniel, 6; Susana, 4; Isaac, 11/12; no property value given.*

### Isaac Fry

- 16 acres improved land valued at $1200.
- Farm machinery valued at $50.
- 1 horse, 2 milch cows, 1 other cattle, 2 swine, valued at $100.
- 45 bu. wheat, 50 bu. Indian corn, 75 bu. oats, 15 bu. Irish potatoes, orchard products valued at $5, 225 lbs. butter, 6 tons hay.
- Animals slaughtered valued at $80.

*Federal census lists Isaac, 53, a master storekeeper; Catherine, 50; Solomon, 25, a storekeeper; and property valued at $1200.*

### George Sindle

- 14 acres improved and 5 acres unimproved land valued at $1500.
- 3 milch cows, 4 swine, valued at $90.
- 46 bu. wheat, 50 bu. Indian corn, 30 bu. oats, 25 bu. Irish potatoes, 175 lbs. butter, 6 tons hay.
- Animals slaughtered valued at $60.

*Federal census lists George, 37, a miller (born in Germany); Lydia, 32; George, 6; Edwin, 3; Moses Miller, 12; Mary Ann, 1; Catherine Hagy, ?2, a widow; and property valued at $1200.*

### William Fry

- 24 acres improved and 2 acres unimproved land valued at $1000.
- 1 milch cow, 1 other cattle, 2 swine, valued at $50.
- 27 bu. wheat, 50 bu. Indian corn, 75 bu. oats, 50 bu. Irish potatoes, 3 bu. sweet potatoes, orchard products valued at $5, 150 lbs. butter, 2 tons hay.
- Animals slaughtered valued at $30.

*Federal census lists William, 22, a day laborer; Elizabeth, 24; Mary Jane, 1; Caroline Fry, 23, serving; and property valued at $1000.*

### William Baussman, tenant

- 50 acres improved and 20 acres unimproved land valued at $3500.
- Farm machinery valued at $150.
- 3 horses, 6 milch cows, 5 other cattle, 8 swine, valued at $300.
- 130 bu. wheat, 40 bu. rye, 130 bu. Indian corn, 200 bu. oats, 25 bu. Irish potatoes, orchard products valued at $10, 300 lbs. butter, 9 tons hay, 6 bu. clover seed, 50 lbs. honey.
- Animals slaughtered valued at $75.

*Federal census lists William, 28; Sarah, 25; Mary Ann, 4; John, 2; Daniel Fankhauser, 12; Abraham, 9; Peter Fankhauser, 20, a day laborer; Peter Fankhauser, 82, a farmer; and property valued at $600.*

### Jesse Martin

- 38 acres improved and 10 acres unimproved land valued at $1600.
- Farm machinery valued at $20.
- 2 horses, 2 milch cows, 1 other cattle, 4 swine, valued at $100.
- 72 bu. wheat, 2 bu. rye, 30 bu. Indian corn, 45 bu. oats, 5 bu. Irish potatoes, 50 lbs. butter, 2 tons hay.
- Animals slaughtered valued at $90.

*Federal census lists Jesse, 23; Elizabeth, 25; Susanna, 3; Catherine, 2; George Wolfskill, 12; and property valued at $1600.*

### Isaac Messner

- 41 acres improved and 4 acres unimproved land valued at $2000.
- Farm machinery valued at $100.
- 4 horses, 5 milch cows, 5 other cattle, 4 sheep, 6 swine, valued at $300.
- 100 bu. wheat, 25 bu. rye, 80 bu. Indian corn, 140 bu. oats, 12 lbs. wool, 30 bu. Irish potatoes, orchard products valued at $5, 300 lbs. butter, 20 tons hay, 25 lbs. flax.
- Animals slaughtered valued at $25.

*Federal census lists Isaac, 51; Harriet, 43; Alexander, 26, a farm laborer; Mary, 19; Elizabeth, 16; Leman, 9; Solomon, 7; Susana, 5; Rebeca, 2; Levi Bixler, 50, a farm laborer; and property valued at $2000.*

### Peter Good
• 30 acres improved and 14 acres unimproved land valued at $2500.
• Farm machinery valued at $100.
• 1 horse, 2 milch cows, 4 other cattle, 3 swine, valued at $200.
• 100 bu. wheat, 150 bu. Indian corn, 150 bu. oats, 20 bu. Irish potatoes, orchard products valued at $5, 150 lbs. butter, 6 tons hay.
• Animals slaughtered valued at $75.
*Federal census lists Peter, 36, a master miller; Hannah, 33; Mary Ann, 9; Leman, 7; Emaline, 4; Moses Wengler, 20, a miller; and property valued at $7500.*

### John Lessley
• 35 acres improved and 5 acres unimproved land valued at $3000.
• Farm machinery valued at $40.
• 3 horses, 4 milch cows, 5 other cattle, 4 swine, valued at $250.
• 151 bu. wheat, 75 bu. Indian corn, 100 bu. oats, 10 bu. Irish potatoes, 300 lbs. butter, 8 tons hay.
• Animals slaughtered valued at $75.
*Federal census lists John, 65; Susanna, 48; Elizabeth, 24; William, 18, a farm laborer; Samuel Fankhauser, 11; and property valued at $3000.*

### Richard Fichthorn
• 28 acres improved and 3 acres unimproved land valued at $1200.
• Farm machinery valued at $30.
• 1 horse, 2 milch cows, 2 other cattle, 3 swine, valued at $100.
• 60 bu. wheat, 60 bu. Indian corn, 60 bu. oats, 30 bu. Irish potatoes, 300 lbs. butter, 10 tons hay.
• Animals slaughtered valued at $52.
*Federal census lists Richard, 35, a hatter; Mary, 29; George, 5; Anna, 3; Cyrus, 3/12; and property valued at $1200.*

### Joel Kint
• 50 acres improved and 10 acres unimproved land valued at $2000.
• Farm machinery valued at $40.
• 2 horses, 5 milch cows, 7 other cattle, 3 swine, valued at $300.
• 20 bu. wheat, 80 bu. rye, 80 bu. Indian corn, 130 bu. oats, 10 bu. Irish potatoes, 400 lbs. butter, 5 tons hay.
• Animals slaughtered valued at $67.
*Federal census lists Joel, 36; Leha, 35; and property valued at $2000.*

### Jacob Lessley
• 75 acres improved and 45 acres unimproved land valued at $6000.
• Farm machinery valued at $100.
• 4 horses, 7 milch cows, 9 other cattle, 2 sheep, 6 swine, valued at $500.
• 275 bu. wheat, 15 bu. rye, 200 bu. Indian corn, 250 bu. oats, 6 lbs. wool, 50 bu. Irish potatoes, 5 bu. sweet potatoes, 500 lbs. butter, 30 tons hay, 1 bu. other grasses, 1 lb. hops.
• Animals slaughtered valued at $130.
*Federal census lists Jacob, 35; Hannah, 43; Jacob, 21, a farm laborer; John, 18, a farm laborer; Joseph, 14, a farm laborer; Henry, 12; Benjamin, 9; Jeramiah, 2; Henrietta, 16; Sophiah, 10; Catherine, 8; Margaret, 6; Susanna, 5; Benjamin, 85; and property valued at $6000.*

### Samuel Fankhauser
• 75 acres improved and 25 acres unimproved land valued at $2000.
• Farm machinery valued at $40.
• 5 horses, 4 milch cows, 6 other cattle, 6 sheep, 9 swine, valued at $200.
• 100 bu. wheat, 10 bu. rye, 100 bu. Indian corn, 150 bu. oats, 20 lbs. wool, 12 bu. Irish potatoes, 200 lbs. butter, 7 tons hay, 1 bu. clover seed, 1 bu. other grass seed.
• Animals slaughtered valued at $75.
*Federal census lists Samuel, 48; Elizabeth, 45; Chambers, 18, a day laborer; Albert, 16, a farm laborer; Catherine, 13; Anna, 9; Samuel, 7; John, 5; and property valued at $2000.*

### Jacob Bieher
• 26 acres improved and 5 acres unimproved land valued at $2400.
• Farm machinery valued at $50.
• 2 horses, 2 milch cows, 3 other cattle, 2 sheep, 3 swine, valued at $150.
• 108 bu. wheat, 20 bu. rye, 150 bu. Indian corn, 75 bu. oats, 10 lbs. wool, 25 bu. Irish potatoes, 200 lbs. butter, 6 tons hay, 1 bu. clover seed, 1 bu. other grass seed.
• Animals slaughtered valued at $97.
*Federal census lists Jacob, 42; Sarah, 36; Jeremiah, 14; Rebecca, 12; Catharine, 7; John, 5; Lydia, 3; Susanna, 1; and property valued at $2400.*

### Jacob Whitman

- 60 acres improved and 20 acres unimproved land valued at $2400.
- Farm machinery valued at $40.
- 2 horses, 2 asses and mules, 4 milch cows, 6 swine, valued at $200.
- 50 bu. wheat, 40 bu. rye, 150 bu. Indian corn, 100 bu. oats, 800 lbs. tobacco, 40 bu. Irish potatoes, 5 bu. buckwheat, 100 lbs. butter, 6 tons hay.
- Animals slaughtered valued at $75.

*Federal census lists Jacob, 44; Anna, 39; Henry 18, a farm laborer; Hitty Ann, 11; Eliza, 1; Cyrus Fry, 14; and property valued at $2400.*

### John Lorah, tenant

- 50 acres improved and 25 acres unimproved land valued at $2000.
- Farm machinery valued at $40.
- 2 horses, 2 milch cows, 3 other cattle, 1 swine, valued at $100.
- 25 bu. wheat, 9 bu. rye, 60 bu. Indian corn, 50 bu. oats, 30 bu. Irish potatoes, 100 lbs. butter, 7 tons hay.
- Animals slaughtered valued at $45.

*Federal census lists David Bixler, 51; Susanna Bixler, 28, serving, and property valued at $2000. John Lorah, 24, a day laborer; Mary Conney, 15; David Lorah, 2; John, 4/12.*

### David McColm

- 69 acres improved and 10 acres unimproved land valued at $3000.
- Farm machinery valued at $200.
- 2 horses, 4 milch cows, 8 other cattle, 4 swine, valued at $400.
- 125 bu. wheat, 300 bu. Indian corn, 250 bu. oats, 50 bu. Irish potatoes, orchard products valued at $5, 350 lbs. butter, 25 tons hay, 10 bu. clover seed, 2 bu. other grass seed.
- Animals slaughtered valued at $75.

*Federal census lists David, 39; Lydia, 31; Mary Ann, 5; Sarah, 3; Ruben Fitch, 12; Peter Weaver, 56, a day laborer, and property valued at $3000.*

### Peter Lineniger

- 17 acres improved and 3 acres unimproved land valued at $1200.
- Farm machinery valued at $40.
- 1 horse, 4 milch cows, 1 other cattle, 6 swine, valued at $150.
- 64 bu. wheat, 2 bu. rye, 50 bu. Indian corn, 8 bu. Irish potatoes, 200 lbs. butter, 5 tons hay, 3 bu. clover seed.
- Animals slaughtered valued at $52.

*Federal census lists Peter, 43, a master shoemaker; Catherine, 40; Elizabeth, 16; Joseph, 14; Hiram, 12; Emme, 10; Mary, 8; Catherine, 6; Peter, 4; Sarah, 1; and property valued at 1200.*

### Samuel Becker, tenant

- 70 acres improved and 30 acres unimproved land valued at $4000.
- Farm machinery valued at $100.
- 4 horses, 5 milch cows, 10 other cattle, 3 swine, valued at $400.
- 185 bu. wheat, 10 bu. rye, 200 bu. Indian corn, 400 bu. oats, 30 bu. Irish potatoes, orchard products valued at $15, 350 lbs. butter, 20 tons hay, 5 bu. clover seed, 3 bu. other grass seed.
- Animals slaughtered valued at $52.

*Federal census lists Samuel, 24; Mary, 25; David, 3; Mary Ann Wolfskill, 16, serving; Israel White, 16, a farm laborer. No property value given.*

### Henry Fry

- 35 acres improved and 40 acres unimproved land valued at $3000.
- Farm machinery valued at $150.
- 2 horses, 6 milch cows, 5 swine, valued at $350.
- 70 bu. wheat, 30 bu. Irish potatoes, 7 bu. sweet potatoes, 550 lbs. butter, 12 tons hay, 10 lbs. honey.
- Animals slaughtered valued at $112.

*Federal census lists Henry, 45; Sarah, 39; Eliza, 17; Susanna, 15; Henry, 6; and property valued at $7000.*

### Ruben Shober, tenant

- 30 acres improved and 14 acres unimproved land valued at $3000.
- Farm machinery valued at $60.
- 2 horses, 4 milch cows, 2 other cattle, 3 swine, valued at $300.
- 85 bu. wheat, 130 bu. Indian corn, 130 bu. oats, 25 bu. Irish potatoes, 2 bu. sweet potatoes, orchard products valued at $6, 260 lbs. butter, 15 tons hay, 4 bu. clover seed.
- Animals slaughtered valued at $97.

*Federal census lists Ruben, 33, a master miller; Margaret, 31; Leha Kinsey, 17, serving; Albert Rupp, 25, a miller.*

### Henry Rupp
- 26 acres improved and 3 acres unimproved land valued at $1800.
- Farm machinery valued at $35.
- 2 horses, 2 milch cows, 2 other cattle, 1 swine, valued at $160.
- 47 bu. wheat, 10 bu. rye, 75 bu. Indian corn, 60 bu. oats, 20 bu. Irish potatoes, 100 lbs. butter, 5 tons hay.
- Animals slaughtered valued at $45.

*Federal census lists Henry, 45, a day laborer; Mary, 50; Henry, 21, a shoemaker; Elizabeth, 19, serving; William, 17, a day laborer; Eliza, 14; George, 12; Rebecca, 10; and property valued at $1800.*

### Levi Laush, tenant
- 26 acres improved and 45 acres unimproved land valued at $1200.
- Farm machinery valued at $100.
- 2 horses, 3 milch cows, 1 other cattle, 3 sheep, 3 swine, valued at $200.
- 45 bu. wheat, 50 bu. Indian corn, 50 bu. oats, 12 lbs. wool, 20 bu. Irish potatoes, 100 lbs. butter, 5 tons hay, 3 bu. other grass seed.
- Animals slaughtered valued at $95.

*Federal census lists Levi, 43; Lydia, 18, serving; Maria, 13; Anna Eliza, 10; John Mooring, 15, a farm laborer; John Laush, 73, owner of property valued at $1200.*

### Peter Hoffman
- 100 acres improved and 14 acres unimproved land valued at $2500.
- Farm machinery valued at $100.
- 4 horses, 4 milch cows, 10 other cattle, 3 sheep, 4 swine, valued at $275.
- 70 bu. wheat, 5 bu. rye, 40 bu. Indian corn, 70 bu. oats, 9 lbs. wool, 20 bu. Irish potatoes, orchard products valued at $5, 125 lbs. butter, 8 tons hay, 1 bu. other grass seed.
- Animals slaughtered valued at $67.

*Federal census lists Peter, 53; Christianna, 47; Lydia, 18; Caroline, 16; Israel, 14; Isaac, 8; and property valued at $2500.*

### William Winehold
- 60 acres improved and 19 acres unimproved land valued at $1600.
- Farm machinery valued at $100.
- 3 horses, 4 milch cows, 4 other cattle, 7 sheep, 4 swine, valued at $300.
- 60 bu. wheat, 20 bu. rye, 100 bu. Indian corn, 60 bu. oats, 15 lbs. wool, 40 bu. Irish potatoes, orchard products valued at $10, 100 lbs. butter, 10 tons hay, 2 bu. clover seed, 10 lbs. honey.
- Animals slaughtered valued at $60.

*Federal census lists William, 62; Maria, 56; Kassey, 22; John, 20, a farm laborer; Jacob, 12; William, 28, a day laborer; and property valued at $1600.*

### Elizabeth Coldren
- 60 acres improved and 20 acres unimproved land valued at $2500.
- Farm machinery valued at $50.
- 3 horses, 4 milch cows, 6 other cattle, 4 swine, valued at $200.
- 60 bu. wheat, 25 bu. rye, 150 bu. Indian corn, 100 bu. oats, 50 bu. Irish potatoes, 400 lbs. butter, 12 tons hay, 2 bu. clover seed.
- Animals slaughtered valued at $90.

*Federal census lists Elizabeth, 55, a widow; Isaac, 33, a day laborer; Elias, 23, a shoemaker; Jacob, 21, a day laborer; Levi, 18, a day laborer; Edward, 15, a day laborer; Mary, 11; Elizabeth, 8; Eave Weaver, 79, a widow; and property valued at $2500.*

### Peter Marks
- 85 acres improved and 15 acres unimproved land valued at $2600.
- Farm machinery valued at $110.
- 3 horses, 4 milch cows, 1 other cattle, 5 swine, valued at $400.
- 20 bu. wheat, 20 bu. rye, 150 bu. Indian corn, 200 bu. oats, 75 bu. Irish potatoes, 39 bu. buckwheat, orchard products valued at $5, 250 lbs. butter, 8 tons hay, 3 bu. clover seed, 5 bu. other grass seed.
- Animals slaughtered valued at $67.

*Federal census lists Peter, 29; Emme, 30; Ellen, 6; Jefferson, 5; Franklin, 3; Edward, 1; Peter Cole, 16, farm laborer; and property valued at $3200.*

### Simon Kindt
- 40 acres improved land valued at $1000.
- Farm machinery valued at $25.
- 2 horses, 4 milch cows, 3 other cattle, 1 sheep, 3 swine, valued at $200.
- 25 bu. wheat, 12 bu. rye, 45 bu. Indian corn, 40 bu. oats, 4 lbs. wool, 10 bu. Irish potatoes, 80 lbs. butter, 5 tons hay, 1 bu. clover seed.
- Animals slaughtered valued at $67.

*Federal census lists Simon, 43; Mary, 43; Sarah, 9; Manda, 5; and property valued at $1000.*

### Samuel Pannybecker
- 50 acres improved and 7 acres unimproved land valued at $1800.
- Farm machinery valued at $20.
- 2 horses, 3 milch cows, 4 other cattle, 3 swine, valued at $200.
- 75 bu. wheat, 9 bu. rye, 150 bu. Indian corn, 150 bu. oats, 25 bu. Irish potatoes, 400 lbs. butter, 10 tons hay.
- Animals slaughtered valued at $90.

*Federal census lists Samuel, 66; Elizabeth Haven, 53, serving; Benawell Pannybecker, 33, a farm laborer; Catherine, 26; and property valued at $1800.*

### William Coldran
- 80 acres improved and 10 acres unimproved land valued at $3000.
- Farm machinery valued at $100.
- 5 horses, 5 milch cows, 7 other cattle, 2 sheep, 7 swine, valued at $600.
- 135 bu. wheat, 100 bu. rye, 200 bu. Indian corn, 200 bu. oats, 50 bu. Irish potatoes, 16 bu. buckwheat, 400 lbs. butter, 20 tons hay.
- Animals slaughtered valued at $175.

*Federal census lists William, 39; Lydia, 38; A. Biram, 14; Owen, 11; Nathaniel, 7; William, 4; David, 11/12; and property valued at $3000.*

### Abraham Trostle
- 100 acres improved and 15 acres unimproved land valued at $3000.
- Farm machinery valued at $30.
- 3 horses, 6 milch cows, 3 other cattle, 12 sheep, 7 swine, valued at $300.
- 85 bu. wheat, 8 bu. rye, 100 bu. Indian corn, 200 bu. oats, 18 lbs. wool, 25 bu. Irish potatoes, 600 lbs. butter, 15 tons hay.
- Animals slaughtered valued at $121.

*Federal census lists Abraham, 76; Mary, 74; Henry, 40, a farm laborer; Catherine, 36; and property valued at $3000.*

### John Fritz
- 56 acres improved and 3 acres unimproved land valued at $2400.
- Farm machinery valued at $50.
- 2 horses, 4 milch cows, 4 other cattle, 5 swine, valued at $300.
- 70 bu. wheat, 25 bu. rye, 100 bu. Indian corn, 100 bu. oats, 30 bu. Irish potatoes, 18 bu. buckwheat, 500 lbs. butter, 10 tons hay.
- Animals slaughtered valued at $75.

*Federal census lists John, 40; Caroline, 36; Sarah, 13; Henry, 11; James, 8; Manda, 5; Mary Anna, 1/12; and property valued at $2100.*

### Samuel Leisy, tenant
- 72 acres improved and 10 acres unimproved land valued at $2000.
- Farm machinery valued at $50.
- 2 horses, 2 milch cows, 6 other cattle, 2 swine, valued at $200.
- 60 bu. wheat, 4 bu. rye, 50 bu. Indian corn, 130 bu. oats, 10 bu. Irish potatoes, 200 lbs. butter, 9 tons hay.
- Animals slaughtered valued at $52.

*Federal census lists Samuel, 40; Mary, 36; Ruben, 18, a farm laborer (twin); Elizabeth, 18 (twin); John, 10; Anna, 5; Mary, 11/12. No property value given.*

### Joseph L. Zerber
- 34 acres improved and 6 acres unimproved land valued at $1800.
- Farm machinery valued at $50.
- 1 horse, 3 milch cows, 2 other cattle, valued at $150.
- 50 bu. wheat, 20 bu. rye, 140 bu. Indian corn, 125 bu. oats, 20 bu. Irish potatoes, 100 lbs. butter, 5 tons hay.
- Animals slaughtered valued at $40.

*Federal census lists Joseph, 37, a carpenter; Susanna, 40; John, 9; Jacob, 7; Lydia, 6; Daniel, 2; Joseph, 11/12; John Dauthing, carpenter's apprentice; and property valued at $800.*

### Jacob Corral
- 20 acres improved land valued at $1000.
- Farm machinery valued at $50.
- 1 horse, 1 milch cow, 3 swine, valued at $125.
- 40 bu. wheat, 25 bu. rye, 50 bu. Indian corn, 30 bu. oats, 20 bu. Irish potatoes, 400 lbs. butter, 10 tons hay, 3 bu. clover seed.
- Animals slaughtered valued at $60.

*Federal census lists Jacob, 52, born in Germany; Catherine, 57; and property valued at $1000.*

## John Keiffer

- 50 acres improved and 20 acres unimproved land valued at $2000.
- Farm machinery valued at $100.
- 2 horses, 5 milch cows, 6 other cattle, 5 swine, valued at $200.
- 45 bu. wheat, 15 bu. rye, 100 bu. Indian corn, 100 bu. oats, 20 bu. Irish potatoes, 300 lbs. butter, 10 tons hay.
- Animals slaughtered valued at $112.

*Federal census lists John, 57; Catherine, 48; Catherine Keiffer, 21, serving; Henry Keiffer, 19, a day laborer; and property valued at $2000. Henry Becker, 28, tenant; Caroline Becker, 27; John, 4; and property valued at $1200.*

## William Echternach

- 93 acres improved and 5 acres unimproved land valued at $2700.
- Farm machinery valued at $150.
- 4 horses, 4 milch cows, 14 other cattle, 4 swine, valued at $400.
- 100 bu. wheat, 17 bu. rye, 150 bu. Indian corn, 120 bu. oats, 25 bu. Irish potatoes, orchard products valued at $10, 400 lbs. butter, 20 tons hay, 1 bu. clover seed.
- Animals slaughtered valued at $90.

*Federal census lists William, 34; Caroline, 29; Albert, 4; Sophia, 6; Justus, 2; Mary, 1; Eliza Zinn, 19, serving; John Brendle, 19, a farm laborer; and property valued at $2700.*

## Henry Slabach

- 100 acres improved and 15 acres unimproved land valued at $3000.
- Farm machinery valued at $100.
- 5 horses, 9 milch cows, 9 other cattle, 6 sheep, 12 swine, valued at $400.
- 100 bu. wheat, 40 bu. rye, 100 bu. Indian corn, 100 bu. oats, 15 lbs. wool, 100 bu. Irish potatoes, 400 lbs. butter, 40 tons hay, 20 lbs. flax, 1 bu. flaxseed.
- Animals slaughtered valued at $75.

*Federal census lists Henry, 65; Lydia, 59; Henry, 30, a farm laborer; Samuel, 20, a farm laborer; Maria, 24; Lydia, 22; John, 17, a day laborer; Sarah, 9; and property valued at $3000.*

## Michael Fry

- 46 acres improved and 10 acres unimproved land valued at $3000.
- Farm machinery valued at $50.
- 2 horses, 5 milch cows, 4 other cattle, 1 sheep, 5 swine, valued at $300.
- 100 bu. wheat, 25 bu. rye, 100 bu. Indian corn, 300 bu. oats, 5 lbs. wool, 50 bu. Irish potatoes, orchard products valued at $10, 300 lbs. butter, 10 tons hay, 3 bu. clover seed.
- Animals slaughtered valued at $75.

*Federal census lists Michael, 53; Lydia, 50; Franklin, 24, a cooper; John, 10; Maranda Fry, 16, serving.*

## Joseph Brendle

- 65 acres improved and 26 acres unimproved land valued at $3500.
- Farm machinery valued at $100.
- 6 horses, 6 milch cows, 4 other cattle, 4 sheep, 4 swine, valued at $600.
- 100 bu. wheat, 50 bu. rye, 100 bu. Indian corn, 200 bu. oats, 10 lbs. wool, 40 bu. Irish potatoes, 2 bu. sweet potatoes, orchard products valued at $8, 600 lbs. butter, 30 tons hay, 2 bu. clover seed.
- Animals slaughtered valued at $97.

*Joseph misidentified as Brendle on the agriculture census. He is listed as Joseph Garman on the federal census (see page 48).*

## William Vanneida

- 40 acres improved and 11 acres unimproved land valued at $4000.
- Farm machinery valued at $150.
- 1 horse, 5 milch cows, 3 other cattle, 4 swine, valued at $300.
- 100 bu. wheat, 75 bu. Indian corn, 140 bu. oats, 8 lbs. wool, 40 bu. Irish potatoes, 300 lbs. butter, 20 tons hay, 1 bu. clover seed.
- Animals slaughtered valued at $270.

*Federal census lists William, 39; Eliza, 36; Martha, 13; Monroe, 12; James, 10; Augustus, 7; William, 4; Elizabeth Good, 21, serving; and property valued at $14,000.*

## Joel Brossman

- 70 acres improved and 30 acres unimproved land valued at $5000.
- Farm machinery valued at $150.
- 3 horses, 3 asses and mules, 6 milch cows, 9 other cattle, 5 sheep, 4 swine, valued at $500.
- 200 bu. wheat, 15 bu. rye, 400 bu. Indian corn, 300 bu. oats, 12 lbs. wool, 100 bu. Irish potatoes, 200 lbs. butter. 30 tons hay, 2 bu. clover seed, 2 bu. other grass seed.
- Animals slaughtered valued at $112.

*Federal census lists Joel, 44; Mary, 42; Henry, 20, a farm laborer; Caroline, 18; Barbara, 12; and property valued at $6000.*

### William Reninger
• 40 acres improved and 20 acres unimproved land valued at $2700.
• Farm machinery valued at $150.
• 3 horses, 6 milch cows, 4 other cattle, 9 swine, valued at $225.
• 125 bu. wheat, 25 bu. rye, 300 bu. Indian corn, 250 bu. oats, 50 bu. Irish potatoes, 7 bu. sweet potatoes, orchard products valued at $10, 500 lbs. butter, 15 tons hay, 1 bu. other grass seed.
• Animals slaughtered valued at $112.

*Federal census lists William, 35; Susanna, 34; Catherine, 9; William, 8; Daniel, 6; Emaline, 3; Cyrus, 1. No property value given.*

### Jacob Laush
• 20 acres improved and 10 acres unimproved land valued at $1000.
• Farm machinery valued at $50.
• 1 horse, 2 milch cows, 1 other cattle, 2 swine, valued at $125.
• 25 bu. wheat, 7 bu. rye, 30 bu. Indian corn, 30 bu. oats, 25 bu. Irish potatoes, 150 lbs. butter, 3 tons hay.
• Animals slaughtered valued at $52.

*Federal census lists Jacob, 56, a master shoemaker; Nancy, 50; and property valued at $1000.*

### Jacob Slabach
• 63 acres improved and 4 acres unimproved land valued at $2000.
• Farm machinery valued at $50.
• 2 horses, 5 milch cows, 4 other cattle, 8 swine, valued at $300.
• 30 bu. wheat, 35 bu. rye, 80 bu. Indian corn, 150 bu. oats, 50 bu. Irish potatoes, 2 bu. sweet potatoes, 300 lbs. butter, 12 tons hay.
• Animals slaughtered valued at $82.

*Federal census lists Jacob, 35; Susanna, 47; Hetty Ann, 10.*

### Christian Good
• 50 acres improved and 6 acres unimproved land valued at 1200.
• Farm machinery valued at $10.
• 1 horse, 2 milch cows, 4 other cattle, 3 swine, valued at $100.
• 20 bu. wheat, 30 bu. rye, 70 bu. Indian corn, 70 bu. oats, 40 bu. Irish potatoes, 6 bu. sweet potatoes, 50 lbs butter, 2 tons hay, 2 bu. clover seed, 10 lbs. flax, 1 bu. flaxseed.
• Animals slaughtered valued at $45.

*Federal census lists Christian, 59; Susanna, 54; Lydia, 18; Samuel, 16; John, 10; and property valued at $1200.*

### Jones Linebach
• 42 acres improved and 3 acres unimproved land valued at $1600.
• Farm machinery valued at $100.
• 2 horses, 3 milch cows, 4 other cattle, 5 swine, valued at $300.
• 20 bu. wheat, 15 bu. rye, 50 bu. Indian corn, 50 bu. oats, 40 bu. Irish potatoes, 100 lbs. butter, 12 tons hay, 1 bu. clover seed.
• Animals slaughtered valued at $90.

*Federal census lists Jones, 36; Esther, 32; Isaac, 9; John, 6; Anna, 2; and property valued at $1600.*

### Peter Steffey
• 30 acres improved land valued at $1600.
• Farm machinery valued at $10.
• 1 horse, 2 milch cows, 2 other cattle, 3 swine, valued at $100.
• 20 bu. wheat, 20 bu. rye, 20 bu. Indian corn, 25 bu. Irish potatoes, 200 lbs. butter, 4 tons hay.
• Animals slaughtered valued at $24.

*Federal census lists Peter, 65; Anna, 21; Emanda, 15; and property valued at $1600.*

### Samuel Coldran, tenant
• 35 acres improved and 5 acres unimproved land valued at $2800.
• Farm machinery valued at $45.
• 1 horse, 3 milch cows, 1 other cattle, 2 swine, valued at $90.
• 50 bu. wheat, 20 bu. rye, 50 bu. Indian corn, 120 bu. oats, 40 bu. Irish potatoes, 2 bu. sweet potatoes, 200 lbs. butter, 8 tons hay.
• Animals slaughtered valued at $37.

*Federal census lists Samuel, 25, a farm laborer; Elizabeth, 23; John, 2; Laman, 9/12; Sarah Redcay, 12, serving. No property value given.*

## Daniel Oberholser

- 100 acres improved and 29 acres unimproved land valued at $6000.
- Farm machinery valued at $250.
- 4 horses, 6 milch cows, 9 other cattle, 13 swine, valued at $500.
- 81 bu. wheat, 54 bu. rye, 200 bu. Indian corn, 300 bu. oats, 100 bu. Irish potatoes, 22 bu. buckwheat, 500 lbs. butter, 15 tons hay, 7 bu. clover seed.
- Animals slaughtered valued at $90.

*Federal census lists Daniel, 52; Sarah, 45; Susanna, 19; John, 13; Esther Oberholser, 54, serving; and property valued at $6000.*

## Henry Gring

- 30 acres improved and 110 acres unimproved land valued at $2000.
- Farm machinery valued at $40.
- 1 horse, 7 milch cows, 7 other cattle, 15 sheep, 3 swine, valued at $300.
- 20 bu. wheat, 8 bu. rye, 20 bu. Indian corn, 27 bu. oats, 30 lbs. wool, 50 bu. Irish potatoes, 3 bu. sweet potatoes, 400 lbs. butter, 12 tons hay.
- Animals slaughtered valued at $112.

*Federal census lists Henry, 49; Elizabeth, 49; Susanna, 19; Allison, 16, male farm laborer; John, 13; Elizabeth, 8.*

## John Frees

- 118 acres improved land valued at $3000.
- Farm machinery valued at $100.
- 4 horses, 13 milch cows, 9 other cattle, 18 swine, valued at $800.
- 200 bu. wheat, 30 bu. rye, 150 bu. Indian corn, 300 bu. oats, 30 bu. Irish potatoes, 2 bu. sweet potatoes, 30 bu. buckwheat, 100 lbs. butter, 25 tons hay, 1 bu. clover seed, 2 bu. other grass seed.
- Animals slaughtered valued at $97.

*Federal census lists John, 65; Mary, 43; Susanna, 21; Rebecca, 19; John, 17; Mary Ann, 15; Kassey Ann, 13; Hannah, 11; Sarah, 9; Ellen, 4; Israel, 2; Emma, 6/12; John Sabo, 24, a day laborer; Maryann Sabo, 4/12; and property valued at $3000.*

## John Steffey

- 17 acres improved and 3 acres unimproved land valued at $1300.
- Farm machinery valued at $50.
- 2 horses, 3 milch cows, 1 other cattle, 4 swine, valued at $200.
- 18 bu. wheat, 30 bu. rye, 30 bu. Indian corn, 23 bu. oats, 20 bu. Irish potatoes, 300 lbs. butter, 5 tons hay.
- Animals slaughtered valued at $60.

*Federal census lists John, 38; Susanna, 36; Kassey, 15; Susanna, 9; Maria, 7; and property valued at $1300.*

## Elias Good

- 50 acres improved and 25 acres unimproved land valued at $3000.
- Farm machinery valued at $50.
- 2 horses, 4 milch cows, 6 other cattle, 4 swine, valued at $300.
- 60 bu. wheat, 25 bu. rye, 75 bu. Indian corn, 100 bu. oats, 25 bu. Irish potatoes, 1 bu. sweet potatoes, 200 lbs. butter, 6 tons hay, 1 bu. clover seed, 1 bu. other grass seed, 12 lbs. noney.
- Animals slaughtered valued at $82.

*Federal census lists Elias, 57; Franey, 55; Mary Shuvert(?), 14; William Good, 15, a farm laborer; and property valued at $3000.*

## John Steffey

- 43 acres improved and 12 acres unimproved land valued at $4000.
- Farm machinery valued at $100.
- 2 horses, 5 milch cows, 6 other cattle, 7 swine, valued at $300.
- 50 bu. wheat, 18 bu. rye, 50 bu. Indian corn, 100 bu. oats, 25 bu. Irish potatoes, 500 lbs. butter, 12 tons hay, 2 bu. clover seed, 1 bu. other grass seed.
- Animals slaughtered valued at $75.

*Federal census lists John, 74; Eave, 76; Lydia, 30; Nancy, 28; and property valued at $4000. Also, Levi Steffey, 48, a farm laborer; Rachel, 43; William, 18, a day laborer, and property valued at $600.*

## Isaac Kessler

- 30 acres improved and 55 acres unimproved land valued at $3000.
- Farm machinery valued at $75.
- 3 horses, 6 milch cows, 4 other cattle, 6 sheep, 11 swine, valued at $300.
- 70 bu. wheat, 50 bu. rye, 50 bu. Indian corn, 100 bu. oats, 20 lbs. wool, 30 bu. Irish potatoes, orchard products valued at $10, 300 lbs. butter, 20 tons hay, 1 bu. clover seed, 1 bu. other grass seed.
- Animals slaughtered valued at $75.

*Federal census lists Isaac, 46; Catherine, 46; Margaret, 14; Daniel, 10; Isaac, 6; Adam, 12; and property valued at $3000.*

**Benjamin Kessler**
• 50 acres improved and 30 acres unimproved land valued at $2000.
• Farm machinery valued at $30.
• 2 horses, 4 milch cows, 4 other cattle, 1 sheep, 4 swine, valued at $400.
• 50 bu. wheat, 40 bu. rye, 50 bu. Indian corn, 100 bu. oats, 3 lbs. wool, 50 bu. Irish potatoes, 200 lbs. butter, 12 tons hay, 3 bu. clover seed, 1 bu. other grass seed.
• Animals slaughtered valued at $75.
*Federal census lists Benjamin, 57; Susana, 51; Mary, 19; Moses White, 14; and property valued at $2000.*

**John Wise**
• 40 acres improved and 30 acres unimproved land valued at $1500.
• Farm machinery valued at $40.
• 2 horses, 5 milch cows, 2 other cattle, 4 sheep, 7 swine, valued at $150.
• 40 bu. wheat, 15 bu. rye, 100 bu. Indian corn, 75 bu. oats, 10 lbs. wool, 75 bu. Irish potatoes, orchard products valued at $15, 200 lbs. butter, 6 tons hay, 3 bu. clover seed, 10 lbs. honey.
• Animals slaughtered valued at $75.
*Federal census lists John, 54; Catherine, 43; Maria, 13; Isaac, 9; Samuel, 7; Susanna, 4; and property valued at $1500.*

**Ulrich Burkholder**
• 20 acres improved and 4 acres unimproved land valued at $900.
• Farm machinery valued at $25.
• 1 horse, 3 milch cows, 1 other cattle, 4 swine, valued at $100.
• 40 bu. wheat, 6 bu. rye, 50 bu. Indian corn, 70 bu. oats, 50 bu. Irish potatoes, orchard products valued at $5, 100 lbs. butter, 4 tons hay.
• Animals slaughtered valued at $82.
*Federal census lists Ulrich, 61; Nancy, 61; and property valued at $900. Also, Esais, 26; Sarah, 22; Caroline, 2; Franklin, 1; and property valued at $1050.*

**Isaac Burkholder**
• 12 acres improved and 9 acres unimproved land valued at $1000.
• Farm machinery valued at $12.
• 2 milch cows, 3 swine, valued at $60.
• 16 bu. wheat, 18 bu. rye, 30 bu. Indian corn, 25 bu. oats, 30 bu. Irish potatoes, 100 lbs. butter, 4 tons hay.
• Animals slaughtered valued at $42.
*Federal census lists Isaac, 37; Catherine, 39; Maria, 11; Rosina, 6; and property valued at $1000.*

**Peter Good**
• 60 acres improved and 50 acres unimproved land valued at $4000.
• Farm machinery valued at $100.
• 2 horses, 4 milch cows, 6 other cattle, 4 swine, valued at $300.
• 100 bu. wheat, 40 bu. rye, 75 bu. Indian corn, 100 bu. oats, 75 bu. Irish potatoes, 5 bu. sweet potatoes, orchard products valued at $5, 400 lbs. butter, 8 tons hay, 6 bu. clover seed.
• Animals slaughtered valued at $75.
*Federal census lists Peter, 75; Lydia, 73; David Good, 30; Sarah, 27; and property valued at $4000.*

**Jacob Shupp**
• 30 acres improved and 40 acres unimproved land valued at $1200.
• Farm machinery valued at $50.
• 2 horses, 4 milch cows, 4 other cattle, 5 swine, valued at $150.
• 35 bu. wheat, 15 bu. rye, 30 bu. Indian corn, 40 bu. oats, 75 bu. Irish potatoes, 1 bu. sweet potatoes, orchard products valued at $10, 300 lbs. butter, 5 tons hay.
• Animals slaughtered valued at $75.
*Federal census lists Jacob, 56; Sophia, 48; Isaac, 24, a farm laborer; Peter, 21, a farm laborer; Susanna, 9; and property valued at $1200. Also, Peter Good, 88; property valued at $600.*

**Isaac Good**
• 65 acres improved and 5 acres unimproved land valued at $3000.
• Farm machinery valued at $100.
• 5 horses, 5 milch cows, 6 other cattle, 5 sheep, 8 swine, valued at $300.
• 80 bu. wheat, 5 bu. rye, 50 bu. Indian corn, 150 bu. oats, 15 lbs. wool, 25 bu. Irish potatoes, 400 lbs. butter, 15 tons hay, 4 bu. clover seed.
• Animals slaughtered valued at $112.
*Federal census lists Isaac, 37; Mary, 40; Martha, 22; Lydia, 35; Margaret, 75, a widow; Sarah Steffey, 15; and property valued at $2000.*

### John Good
- 45 acres improved and 5 acres unimproved land valued at $2000.
- Farm machinery valued at $50.
- 2 horses, 5 milch cows, 6 other cattle, 7 swine, valued at $200.
- 70 bu. wheat, 25 bu. rye, 50 bu. Indian corn, 100 bu. oats, 25 bu. Irish potatoes, 2 bu. sweet potatoes, 300 lbs. butter, 15 tons hay, 1 bu. clover seed.
- Animals slaughtered valued at $75.

*Federal census lists John, 46; Barbara, 35; Alleson (a male), 5; Aaron, 8; Maria, 4; Christiana Bitzen, 14; and property valued at $1200.*

### Catharine Steffey
- 50 acres improved and 50 acres unimproved land valued at $2800.
- Farm machinery valued at $100.
- 2 horses, 4 milch cows, 1 other cattle, 6 swine, valued at $300.
- 80 bu. wheat, 20 bu. rye, 100 bu. Indian corn, 100 bu. oats, 25 bu. Irish potatoes, 200 lbs. butter, 10 tons hay, 3 bu. clover seed.
- Animals slaughtered valued at $52.

*Federal census lists Catharine, 63, a widow; John Steffey, 35, a farm laborer; Louisa, 28; Margaret, 20; Joseph, 18, a farm laborer; and property valued at $2800.*

### Isaac Kern
- 68 acres improved and 30 acres unimproved land valued at $2000.
- Farm machinery valued at $75.
- 4 horses, 5 milch cows, 7 other cattle, 2 sheep, 6 swine, valued at $200.
- 120 bu. wheat, 150 bu. Indian corn, 150 bu. oats, 10 lbs. wool, 50 bu. Irish potatoes, 300 lbs. butter, 12 tons hay.
- Animals slaughtered valued at $90.

*Federal census lists Isaac, 49; Barbara, 18; Catherine, 16; Elizabeth, 15; Martin, 14; Adam, 13; John, 12; Mary, 8; Susanna, 7; Hettie, 4; Milton, 3; and property valued at $2000.*

### Joseph Beler
- 60 acres improved and 24 acres unimproved land valued at $1000.
- Farm machinery valued at $50.
- 2 horses, 4 milch cows, 6 other cattle, 4 swine, valued at $150.
- 25 bu. wheat, 12 bu. rye, 60 bu. Indian corn, 60 bu. oats, 60 bu. Irish potatoes, 300 lbs. butter, 10 tons hay.
- Animals slaughtered valued at $75.

*Federal census lists Joseph, 58; Elizabeth, 61; Joseph, 16; and property valued at $1000.*

### Samuel Davis
- 40 acres improved and 10 acres unimproved land valued at $900.
- Farm machinery valued at $50.
- 4 horses, 4 milch cows, 5 other cattle, 3 swine, valued at $200.
- 50 bu. wheat, 8 bu. rye, 30 bu. Indian corn, 30 bu. oats, 30 bu. Irish potatoes, 4 bu. buckwheat, orchard products valued at $5, 200 lbs. butter, 5 tons hay.
- Animals slaughtered valued at $52.

*Federal census lists Samuel, 35; Sarah, 40; Elizabeth, 10; Simon, 7; Catherine, 5; Loversa, 2; Clinton, 1; Adam Huet, 15; and property valued at $900.*

### Amos Good
- 68 acres improved and 62 acres unimproved land valued at $3500.
- Farm machinery valued at $100.
- 3 horses, 6 milch cows, 12 other cattle, 4 sheep, 4 swine, valued at $400.
- 100 bu. wheat, 18 bu. rye, 75 bu. Indian corn, 150 bu. oats, 15 lbs. wool, 20 bu. Irish potatoes, 400 lbs. butter, 20 tons hay, 3 bu. clover seed, 1 bu. other grass seed.
- Animals slaughtered valued at $60.

*Federal census lists Amos, 23; Lavina, 20; John Bixler, 12; Elizabeth, 15; Esther, 50, a widow; Catherine, 26; and property valued at $3500.*

### Jones Good
- 52 acres improved and 10 acres unimproved land valued at $2500.
- Farm machinery valued at $100.
- 3 horses, 6 milch cows, 5 other cattle, 5 swine, valued at $300.
- 125 bu. wheat, 30 bu. rye, 100 bu. Indian corn, 140 bu. oats, 30 bu. Irish potatoes, 3 bu. sweet potatoes, 500 lbs. butter, 20 tons hay, 4 bu. clover seed.
- Animals slaughtered valued at $150.

*Federal census lists Jones, 47; Catherine, 45; Joseph, 16, a farm laborer; Maria Good, 19, serving; Mary Good, 43; Anna Garman, 3; and property valued at $2500. Jones, 65; Nancy, 30; Juda, 17; and property valued at $3500.*

### Benjamin Gaman
• 80 acres improved and 28 acres unimproved land valued at $3600.
• Farm machinery valued at $100.
• 5 horses, 4 milch cows, 6 other cattle, 3 sheep, 4 swine, valued at $500.
• 75 bu. wheat, 75 bu. rye, 200 bu. Indian corn, 150 bu. oats, 100 lbs. tobacco, 10 lbs. wool, 100 bu. Irish potatoes, 2 bu. sweet potatoes, orchard products valued at $5, 400 lbs. butter, 15 tons hay, 2 bu. clover seed, 30 lbs. flax, 2 bu. flaxseed.
• Animals slaughtered valued at $60.
*Federal census lists Benjamin, 50; Elizabeth, 47; Solomon, 34, a sagar maker; Sarah, 21, serving; Maria, 18, serving; Israel, 16, a shoemaker; Joseph, 13; Elizabeth, 11; Benjamin, 9; John, 7; Noah, 4; Christian Messner, 78, a farmer; and property valued at $8000.*

### John Schneder
• 24 acres improved and 7 acres unimproved land valued at $1400.
• Farm machinery valued at $50.
• 1 horse, 2 milch cows, 1 other cattle, 3 swine, valued at $200.
• 30 bu. wheat, 20 bu. rye, 50 bu. Indian corn, 35 bu. oats, 100 bu. Irish potatoes, 2 bu. sweet potatoes, orchard products valued at $20, 100 lbs. butter, 20 tons hay.
• Animals slaughtered valued at $112.
*Federal census lists John, 28, a day laborer; Mary, 27; Mary, 4; John, 1; Benjamin Good, 57; Richard Good, 23; and property valued at $1400.*

### Benjamin Musser
• 47 acres improved and 11 acres unimproved land valued at $2000.
• Farm machinery valued at $50.
• 1 horse, 4 milch cows, 9 other cattle, 4 swine, valued at $150.
• 50 bu. wheat, 20 bu. rye, 100 bu. Indian corn, 160 bu. oats, 25 bu. Irish potatoes, orchard products valued at $5, 100 lbs. butter, 4 tons hay, 1 bu. clover seed.
• Animals slaughtered valued at $52.
*Federal census lists Benjamin, 52; Elizabeth, 46; Lydia, 18, serving; Elizabeth, 13; Samuel, 11; and property valued at $2000.*

### John Lorah
• 130 acres improved and 30 acres unimproved land valued at $10,000.
• Farm machinery valued at $200.
• 6 horses, 7 milch cows, 19 other cattle, 14 sheep, 8 swine, valued at $1000.
• 380 bu. wheat, 95 bu. rye, 700 bu. Indian corn, 450 bu. oats, 25 lbs.wool, 100 bu. Irish potatoes, 2 bu. sweet potatoes, orchard products valued at $15, 700 lbs. butter, 35 tons hay, 15 bu. clover seed, 10 lbs. flax, 1 bu. flaxseed.
• Homemade manufactures valued at $20.
• Animals slaughtered valued at $203.
*Federal census lists John, 52; Elizabeth, 49; Jacob, 22, farm laborer; Hannah, 20; Elizabeth, 18; Millia, 16 (female); George, 14; Maria, 10; Anna, 7; and property valued at $10,000.*

### Samuel Stober
• 42 acres improved and 8 acres unimproved land valued at $1200.
• Farm machinery valued at $25.
• 2 horses, 3 milch cows, 2 other cattle, 3 swine, valued at $200.
• 50 bu. wheat, 15 bu. rye, 80 bu. Indian corn, 65 bu. oats, 15 bu. Irish potatoes, 5 bu. sweet potatoes, 300 lbs. butter, 8 tons hay.
• Animals slaughtered valued at $60.
*Federal census lists Samuel, 33; Mary, 35; Elmira, 9; Monroe, 8; Susanna Stober, 72, a widow; and property valued at $1200.*

### Benjamin Keiffer
• 75 acres improved and 25 acres unimproved land valued at $2500.
• Farm machinery valued at $50.
• 4 horses, 6 milch cows, 6 other cattle, 5 swine, valued at $250.
• 60 bu. wheat, 12 bu. rye, 100 bu. Indian corn, 100 bu. oats, 30 bu. Irish potatoes, 600 lbs. butter, 10 tons hay, 1 bu. clover seed, 1 bu. other grass seed.
• Animals slaughtered valued at $105.
*Federal census lists Benjamin, 48; Samuel, 40, a tenant; Susanna, 40; Elizabeth, 7; and property valued at $2500; tenant's value, $150.*

## Isaac Keiffer
- 25 acres improved and 10 acres unimproved land valued at $700.
- 3 milch cows, 2 other cattle, 2 swine, valued at $75.
- 20 bu. wheat, 6 bu. rye, 40 bu. Indian corn, 40 bu. oats, 30 bu. Irish potatoes, 300 lbs. butter, 5 tons hay.
- Animals slaughtered valued at $67.

*Federal census lists Isaac, 56; Elizabeth, 56; Eliza, 16; and property valued at $700.*

## Isaac Roth
- 60 acres improved and 20 acres unimproved land valued at $2000.
- Farm machinery valued at $50.
- 2 horses, 4 milch cows, 7 other cattle, 5 sheep, 6 swine, valued at $200.
- 75 bu. wheat, 5 bu. rye, 50 bu. Indian corn, 100 bu. oats, 10 lbs. wool, 30 bu. Irish potatoes, 400 lbs. butter, 20 tons hay, 3 bu. clover seed, 1 bu. other grass seed, 10 lbs. flax, 1 bu. flaxseed.
- Animals slaughtered valued at $105.

*Federal census lists Isaac, 44; Phianna, 30; Franklin, 16; Elizabeth, 4; Lydia, 3; Emeline, 9/12; Esther Roth, 68, a widow; and property valued at $2000.*

## Catharine Steffey
- 60 acres improved and 40 acres unimproved land valued at $2800.
- Farm machinery valued at $100.
- 2 horses, 4 milch cows, 1 other cattle, 6 swine, valued at $200.
- 80 bu. wheat, 20 bu. rye, 100 bu. Indian corn, 90 bu. oats, 30 bu. Irish potatoes, 100 lbs. butter, 12 tons hay.
- Animals slaughtered valued at $75.

*Federal census lists Catharine, 63, widow; John, 35, a farm laborer; Sausia, 28; Margaret, 20; Joseph, 18, a farm laborer.*

## Samuel Lessley
- 20 acres improved land valued at $2000.
- Farm machinery valued at $50.
- 2 horses, 3 milch cows, 1 other cattle; 4 swine, valued at $200.
- 60 bu. wheat, 150 bu. Indian corn, 125 bu. oats, 20 bu. Irish potatoes, 300 lbs. butter, 6 tons hay.
- Animals slaughtered valued at $105.

*Federal census lists Samuel, 52, a master blacksmith; Margaret, 53; John, 22, a blacksmith; Elizabeth, 19; Solomon, 12; and property valued at $2000.*

## Peter Good
- 56 acres improved and 20 acres unimproved land valued at $4000.
- Farm machinery valued at $52.
- 2 horses, 5 milch cows, 6 other cattle, 4 swine, valued at $300.
- 140 bu. wheat, 250 bu. Indian corn, 200 bu. oats, 20 bu. Irish potatoes, orchard products valued at $5, 400 lbs. butter, 12 tons hay, 1 bu. clover seed.
- Animals slaughtered valued at $112.

*Federal census lists Peter, 53; Elizabeth, 50; Israel, 29, a mason; Peter, 17, a day laborer; and property valued at $4000.*

## John Youse
- 15 acres improved land valued at $1800.
- Farm machinery valued at $50.
- 2 milch cows, 1 other cattle, 4 swine, valued at $50.
- 35 bu. wheat, 90 bu. Indian corn, 40 bu. oats, 20 bu. Irish potatoes, orchard products valued at $5, 200 lbs. butter, 4 tons hay, 2 bu. clover seed.
- Animals slaughtered valued at $75.

*Federal census lists John, 44, a carpenter; Sarah, 43; Sausia, 12; Samuel, 10; Lydia Ann, 8; Catherine, 5; Leman, 1; and property valued at $1800.*

## Jacob Root
- 92 acres improved and 8 acres unimproved land valued at $6000.
- Farm machinery valued at $75.
- 6 horses, 7 milch cows, 7 other cattle, 13 swine, valued at $800.
- 200 bu. wheat, 20 bu. rye, 300 bu. Indian corn, 400 bu. oats, 30 bu. Irish potatoes, 400 lbs. butter, 25 tons hay, 3 bu. clover seed, 1 bu. other grass seed.
- Animals slaughtered valued at $800.

*Federal census lists Jacob, 35; Nancy, 28; Magdalinea, 6; John J., 4; George, 2; Mary, 1; John Martin, 22, a farm laborer; Nancy Moyer, 14, serving.*

## Jacob Coldren
- 17 acres improved land valued at $1000.
- Farm machinery valued at $40.
- 2 horses, 1 milch cow, 3 other cattle, 6 sheep, 14 swine, valued at $150.
- 30 bu. wheat, 50 bu. Indian corn, 50 bu. oats, 12 lbs. wool, 25 bu. Irish potatoes, orchard products valued at $5, 75 lbs. butter, 5 tons hay.
- Animals slaughtered valued at $52.

*Federal census lists Jacob, 40, a day laborer; Lydia, 45; Rudy Killian, 4; Mary Myer, 42; and property valued at $1000. Magdelena, 77, a widow and tenant; Mary Ann Coldren, 10.*

## John Coldran
- 50 acres improved and 15 acres unimproved land valued at $1500.
- Farm machinery valued at $40.
- 4 horses, 5 milch cows, 4 other cattle, 2 sheep, 5 swine, valued at $100.
- 100 bu. wheat, 10 bu. rye, 200 bu. Indian corn, 200 bu. oats, 12 lbs. wool, 40 bu. Irish potatoes, 150 lbs. butter, 10 tons hay, 2 bu. clover seed, 20 lbs. flax, 3 bu. flaxseed.
- Animals slaughtered valued at $75.

*Federal census lists John, 56; Barbara, 50; William, 24, a day laborer; Adam, 21, a day laborer; Ephriam, 19, a day laborer; Lydia, 17; John, 15; Samuel, 13; Henry, 12; Solomon, 10; Jacob, 6; and property valued at $1500.*

## Abraham Sollenberger
- 60 acres improved and 31 acres unimproved land valued at $5000.
- Farm machinery valued at $200.
- 4 horses, 7 milch cows, 14 other cattle, 7 swine, valued at $700.
- 260 bu. wheat, 40 bu. rye, 200 bu. Indian corn, 339 bu. oats, 70 bu. Irish potatoes, 4 bu. sweet potatoes, orchard products valued at $10, 700 lbs. butter, 30 tons hay, 3 bu. clover seed, 2 bu. other grass seed, 5 lbs. flax.
- Animals slaughtered valued at $150.

*Federal census lists Abraham, 58; Nancy, 48; Anna, 22; John, 16; Daniel, 23, a carpenter and Lydia, 21, married in last year; Abraham, 9/12; and property valued at $5000.*

## Daniel Musser
- 100 acres improved and 20 acres unimproved land valued at $5000.
- Farm machinery valued at $30.
- 2 horses, 2 milch cows, 1 other cattle, 2 swine, valued at $100.
- 200 bu. wheat, 50 bu. rye, 200 bu. Indian corn, 200 bu. oats, 30 bu. Irish potatoes, 300 lbs. butter, 12 tons hay, 2 bu. clover seed, 2 bu. other grass seed.
- Animals slaughtered valued at $75.

*Federal census lists Daniel, 62; Elizabeth, 61; Nancy, 25; Daniel, 18, a farm laborer; and property valued at $5000.*

## Peter Becker
- 40 acres improved and 14 acres unimproved land valued at $2000.
- Farm machinery valued at $100.
- 1 horse, 4 milch cows, 5 other cattle, 4 swine, valued at $125.
- 90 bu. wheat, 19 bu. rye, 200 bu. Irish potatoes, 135 bu. oats, 50 bu. Irish potatoes, 4 bu. sweet potatoes, 2 bu. buckwheat, orchard products valued at $10, 300 lbs. butter, 8 tons hay, 4 bu. clover seed, 1 bu. other grass seed, 4 lbs. beeswax, 12 lbs. honey.
- Animals slaughtered valued at $112.

*Federal census lists Peter, 59; Esther, 59.*

## John Brossman
- 23 acres improved and 2 acres unimproved land valued at $2000.
- Farm machinery valued at $50.
- 1 horse, 2 milch cows, 1 other cattle, 4 swine, valued at $150.
- 35 bu. wheat, 10 bu. rye, 50 bu. Indian corn, 30 bu. oats, 20 bu. Irish potatoes, orchard products valued at $10, 200 lbs. butter, 8 tons hay, 1 bu. clover seed, 2 bu. other grass seed.
- Animals slaughtered valued at $60.

*Federal census lists John, 56; Hannah, 55; Rudy, 21, a farm laborer; Linda Stober, 11; and property valued at $5000.*

## Isaac Reninger
- 85 acres improved and 6 acres unimproved land valued at $5900.
- Farm machinery valued at $300.
- 4 horses, 6 milch cows, 8 other cattle, 7 swine, valued at $400.
- 160 bu. wheat, 40 bu. Indian corn, 150 bu. oats, 15 bu. Irish potatoes, 4 bu. sweet potatoes, 200 lbs. butter, 4 tons hay, 2 bu. clover seed, 16 lbs. honey.

• Animals slaughtered valued at $120.

*Federal census lists Isaac, 45; Rebecca, 35; Martin, 18, a farm laborer; Maria, 16; Baltzer, 7/12; and property valued at $5900.*

**John Bauman**

• 57 acres improved and 17 acres unimproved land valued at $5500.
• Farm machinery valued at $150.
• 5 horses, 5 milch cows, 5 other cattle, 4 swine, valued at $500.
• 130 bu. wheat, 25 bu. rye, 120 bu. Indian corn, 150 bu. oats, 30 bu. Irish potatoes, orchard products valued at $5, 400 lbs. butter, 12 tons hay, 1 bu. clover seed.
• Animals slaughtered valued at $105.

*Federal census lists John, 19; Elizabeth, 21; Levina Bauman, 17, serving; Rachel Redcay, 13; and property valued at $5500.*

**Jacob Reninger**

• 34 acres improved land valued at $2600.
• Farm machinery valued at $50.
• 2 horses, 2 milch cows, 2 other cattle, 4 swine, valued at $160.
• 67 bu. wheat, 10 bu. rye, 50 bu. Indian corn, 90 bu. oats, 20 bu. Irish potatoes, 200 lbs. butter, 5 tons hay.
• Animals slaughtered valued at $60.

*Federal census lists Jacob, 48; Catherine, 44; Caroline, 19, a seamstress; Edward, 11; Christian, 8; Hannah, 6; Susanna, 4; Amanda, 7/12; and property valued at $2600.*

**Cyrus Messner**

• 34 acres improved and 2 acres unimproved land valued at $2000.
• Farm machinery valued at $50.
• 2 horses, 5 milch cows, 6 other cattle, 5 swine, valued at $200.
• 50 bu. wheat, 20 bu. rye, 50 bu. Indian corn, 50 bu. oats, 20 bu. Irish potatoes, 200 lbs. butter, 2 tons hay.
• Animals slaughtered valued at $90.

*Federal census lists Cyrus, 45, a carpenter; Lydia, 42; Davis, 21, a carpenter; Tobias, 15; Lydia, 5; Heber, 3; Levi, 2/12; and property valued at $2000.*

**Rudy Killian**

• 85 acres improved and 5 acres unimproved land valued at $3000.
• Farm machinery valued at $100.
• 5 horses, 5 milch cows, 6 other cattle, 1 sheep, 10 swine, valued at $250.
• 350 bu. wheat, 10 bu. rye, 100 bu. Indian corn, 100 bu. oats, 4 lbs. wool, 20 bu. Irish potatoes, 2 bu. sweet potatoes, 300 lbs. butter, 10 tons hay.
• Animals slaughtered valued at $105.

*Federal census lists Rudy, 54; Ellen, 50; Susanna, 25; Sophia, 19; Rudy, 17, a farm laborer; Emanuel, 14; Margaret, 13; Geave Woodher/Weadher, 5; and property valued at $3000.*

**Samuel Winehold**

• 62 acres improved and 10 acres unimproved land valued at $3000.
• Farm machinery valued at $100.
• 3 horses, 4 milch cows, 8 other cattle, 5 sheep, 11 swine, valued at $150.
• 150 bu. wheat, 22 bu. rye, 200 bu. Indian corn, 200 bu. oats, 10 lbs. wool, 30 bu. Irish potatoes, 2 bu. sweet potatoes, 400 lbs. butter, 12 tons hay, 1 bu. clover seed, 1 bu. other grass seed.
• Animals slaughtered valued at $90.

*Federal census lists Samuel, 43; Hannah, 37; Daniel, 17; Samuel, 14; Mary, 12; John, 10; and property valued at $3000.*

**Joseph Good**

• 75 acres improved and 15 acres unimproved land valued at $3500.
• Farm machinery valued at $100.
• 2 horses, 6 milch cows, 7 other cattle, 7 swine, valued at $300.
• 70 bu. wheat, 20 bu. rye, 200 bu. Indian corn, 200 bu. oats, 100 bu. Irish potatoes, 25 bu. buckwheat, orchard products valued at $15, 300 lbs. butter, 12 tons hay, 4 bu. clover seed, 1 bu. other grass seed.
• Animals slaughtered valued at $105.

*Federal census lists Joseph, 25; Anna, 22; Solomon Good, 29, a master shoemaker, Jones Good, 65; Nancy, 30; Juda, 17; and property valued at $3500.*

**John Becker**

• 60 acres improved and 5 acres unimproved land valued at $3000.
• Farm machinery valued at $50.
• 3 horses, 5 milch cows, 8 other cattle, 10 swine, valued at $200.

- 90 bu. wheat, 40 bu. rye, 150 bu. Indian corn, 300 bu. oats, 80 bu. Irish potatoes, 400 lbs. butter, 12 tons hay, 5 bu. clover seed, 2 bu. other grass seed, 25 lbs. flax, 2 bu. flaxseed.
- Animals slaughtered valued at $75.

*Federal census lists John, 61; Peter, 24, a farm laborer; Maria, 17; Rebecca, 16; and property valued at $3000.*

### David W. Bender
- 44 acres improved and 10 acres unimproved land valued at $2000.
- Farm machinery valued at $150.
- 2 horses, 2 milch cows, 4 other cattle, 3 swine, valued at $450.
- 83 bu. wheat, 20 bu. rye, 150 bu. Indian corn, 120 bu. oats, 20 bu. Irish potatoes, 8 bu. buckwheat, orchard products valued at $5, 400 lbs. butter, 10 tons hay.
- Animals slaughtered valued at $90.

*Federal census lists David, 29; Elizabeth, 29; James, 5; Maria, 2; and property valued at 2000.*

### Isaac Bauman
- 40 acres improved and 5 acres unimproved land valued at $2000.
- Farm machinery valued at $50.
- 2 horses, 3 milch cows, 3 other cattle, 3 swine, valued at $200.
- 60 bu. wheat, 10 bu. rye, 160 bu. Indian corn, 150 bu. oats, 40 bu. Irish potatoes, 200 lbs. butter, 12 tons hay.
- Animals slaughtered valued at $60.

*Federal census lists Isaac, 43; Leha, 45; Elias, 20, insane; Susanna, 12; Daniel, 10; Elizabeth, 1; and property valued at $2000.*

### Johnason Good
- 62 acres improved and 8 acres unimproved land valued at $3600.
- Farm machinery valued at $250.
- 2 horses, 2 asses and mules, 4 milch cows, 10 other cattle, 5 swine, valued at $500.
- 180 bu. wheat, 15 bu. rye, 200 bu. Indian corn, 300 bu. oats, 75 bu. Irish potatoes, orchard products valued at $10, 500 lbs. butter, 15 tons hay, 1 bu. clover seed, 1 bu. other grass seed.
- Animals slaughtered valued at $112.

*Federal census lists Johnason, 25; Catherine, 26; Manda, 5; Mary, 3; Moses, 11/12; Samuel Sifried, 15, a farm laborer. No property value given.*

### Daniel Gaman
- 82 acres improved and 8 acres unimproved land valued at $3500.
- Farm machinery valued at $150.
- 3 horses, 3 milch cows, 6 other cattle, 4 swine, valued at $800.
- 90 bu. wheat, 40 bu. rye, 60 bu. Indian corn, 100 bu. oats, 100 bu. Irish potatoes, 24 bu. buckwheat, orchard products valued at $5, 200 lbs. butter, 30 tons hay, 2 bu. clover seed, 1 bu. other grass seed.
- Animals slaughtered valued at $83.

*Federal census lists Daniel, 47; Linda, 38; John Kramer, 8; Lavesa Grosendman, 10; Jeremiah Brendle, 13; and property valued at $3500. Also, Nancy Musselman, 41, a widow; property valued at $3500.*

### Jacob Good
- 35 acres improved and 6 acres unimproved land valued at $2500.
- Farm machinery valued at $100.
- 2 horses, 3 milch cows, 1 other cattle, 3 swine, valued at $350.
- 80 bu. wheat, 30 bu. rye, 120 bu. Indian corn, 150 bu. oats, 50 bu. Irish potatoes, 300 lbs. butter, 10 tons hay, 2 bu. clover seed, 1 bu. other grass seed.
- Animals slaughtered valued at $165.

*Federal census lists Jacob, 48; Rachel, 38; Ellen, 8; William, 6; Caroline, 1; Jacob Good, 25, a farm laborer; and property valued at $4600.*

### Jacob Musselman
- 24 acres improved and 6 acres unimproved land valued at $4000.
- Farm machinery valued at $150.
- 1 horse, 2 milch cows, 3 swine, valued at $200.
- 50 bu. wheat, 80 bu. Indian corn, 60 bu. oats, 60 bu. Irish potatoes, 10 bu. sweet potatoes, 5 bu. buckwheat, 500 lbs. butter, 25 tons hay, 3 bu. clover seed.
- Animals slaughtered valued at $50.

*Possibly Jacob, 23; Elizabeth, 21.*

### Adam Kessler
- 43 acres improved and 7 acres unimproved land valued at $3000.
- Farm machinery valued at $100.

- 3 horses, 4 milch cows, 4 other cattle, 3 sheep, 4 swine, valued at $300.
- 60 bu. wheat, 15 bu. rye, 50 bu. Indian corn, 100 bu. oats, 13 lbs. wool, 50 bu. Irish potatoes, 2 bu. sweet potatoes, orchard products valued at $5, 400 lbs. butter, 12 tons hay, 4 bu. clover seed, 2 bu. other grass seed.
- Homemade manufactures valued at $8.
- Animals slaughtered valued at $75.

*Federal census lists Adam, 60; Rachel, 55; John, 15; Elizabeth, 12; and property valued at $3000. Also, Mary Musselman, 21, a widow; property valued at $1000.*

### John Musselman

- 70 acres improved and 17 acres unimproved land valued at $4500.
- Farm machinery valued at $250.
- 4 horses, 7 milch cows, 6 other cattle, 7 swine, valued at $800.
- 120 bu. wheat, 40 bu. rye, 250 bu. Indian corn, 300 bu. oats, 40 bu. Irish potatoes, 70 lbs. butter, 2 tons hay.
- Animals slaughtered valued at $112.

*Federal census lists John, 50; Nancy, 50; Elizabeth, 30; John, 23, a farm laborer; John, 13; and property valued at $4500.*

### Moses Bauman

- 54 acres improved and 15 acres unimproved land valued at $3700.
- Farm machinery valued at $200.
- 2 horses, 6 milch cows, 4 other cattle, 1 sheep, 5 swine, valued at $250.
- 90 bu. wheat, 50 bu. rye, 150 bu. Indian corn, 240 bu. oats, 3 lbs. wool, 50 bu. Irish potatoes, orchard products valued at $15, 500 lbs. butter, 20 tons hay, 3 bu. clover seed, 1 bu. other grass seed.
- Animals slaughtered valued at $105.

*Federal census lists Moses, 34; Mary, 35; Gurian, 5 (male); Anna, 1; Molly Sitesinger, 13; and property valued at $3700. Daniel Bauman, 64; Elizabeth, 62; and property valued at $400.*

### Jacob Musser

- 66 acres improved and 6 acres unimproved land valued at $3500.
- Farm machinery valued at $100.
- 3 horses, 5 milch cows, 8 other cattle, 5 swine, valued at $400.
- 120 bu. wheat, 20 bu. rye, 125 bu. Indian corn, 175 bu. oats, 30 bu. Irish potatoes, orchard products valued at $5, 300 lbs. butter, 15 tons hay, 2 bu. clover seed, 1 bu. other grass seed.
- Animals slaughtered valued at $134.

*Federal census lists Jacob, 38; Lydia, 43; Joseph, 17, a farm laborer; Elizabeth, 16; Susanna, 12; Solomon, 7; Lydia, 4; Jacob, 2; and property valued at $3500.*

### Henry Vanneida

- 13 acres improved and 27 acres unimproved land valued at $2000.
- Farm machinery valued at $250.
- 2 horses, 4 milch cows, 4 other cattle, 6 swine, valued at $350.
- 95 bu. wheat, 50 bu. oats, 2 bu. peas and beans, 120 bu. Irish potatoes, 4 bu. sweet potatoes, orchard products valued at $5, 400 lbs. butter, 15 tons hay.
- Animals slaughtered valued at $126.

*Federal census lists Henry, 34, a master miller; Sarah, 28; Emaline, 7; Isabela, 2; Israel Moon, 20, a miller; Martin Lorah, 25, a farm laborer; Harriet Artz, 17, serving; and property valued at $11,000.*

### Samuel Good

- 73 acres improved and 20 acres unimproved land valued at $2000.
- Farm machinery valued at $50.
- 2 milch cows, 1 other cattle, 4 swine, valued at $50.
- 20 bu. wheat, 80 bu. oats, 13 bu. Irish potatoes, 100 lbs. butter, 5 tons hay.
- Animals slaughtered valued at $45.

*Federal census lists Samuel, 69; and property valued at $2000.*

### Peter Musser

- 83 acres improved and 11 acres unimproved land valued at $3000.
- Farm machinery valued at $100.
- 3 horses, 5 milch cows, 6 other cattle, 5 swine, valued at $200.
- 140 bu. wheat, 27 bu. rye, 200 bu. Indian corn, 200 bu. oats, 40 bu. Irish potatoes, 300 lbs. butter, 12 tons hay.
- Animals slaughtered valued at $105.

*Federal census lists Peter, 40; Sarah, 40; Franklin, 11; Israel, 9; Mary Ann, 8; and property valued at $3000.*

### Richard Beam

- 112 acres improved land valued at $2000.
- Farm machinery valued at $50.

- 2 horses, 4 milch cows, 10 other cattle, 3 swine, valued at $300.
- 60 bu. wheat, 20 bu. rye, 100 bu. Indian corn, 150 bu. oats, 30 bu. Irish potatoes, 10 bu. sweet potatoes, orchard products valued at $10, 400 lbs. butter, 10 tons hay, 3 bu. clover seed, 10 lbs. flax, 1 bu. flaxseed.
- Animals slaughtered valued at $90.

*Federal census lists Richard, 40; Salome, 41; Susanna, 10; Sarah, 7; Mary, 3; and property valued at $2000.*

### Daniel Palm

- 16 acres improved land valued at $1700.
- 3 milch cows, 1 other cattle, 5 swine, valued at $100.
- 25 bu. wheat, 40 bu. Indian corn, 28 bu. oats, 18 bu. Irish potatoes, 9 bu. sweet potatoes, 300 lbs. butter, 4 tons hay.
- Animals slaughtered valued at $90.

*Federal census lists Daniel, 41, a master blacksmith; Nancy, 41; Jane, 17; Peter, 14; and property valued at $1700.*

### Samuel Beam

- 80 acres improved and 20 acres unimproved land valued at $3000.
- Farm machinery valued at $50.
- 3 horses, 5 milch cows, 7 other cattle, 5 swine, valued at $500.
- 100 bu. wheat, 50 bu. rye, 100 bu. Indian corn, 400 bu. oats, 70 bu. Irish potatoes, 30 bu. buckwheat, orchard products valued at $10, 500 lbs. butter, 20 tons hay, 6 bu. clover seed.
- Animals slaughtered valued at $150.

*Federal census lists Samuel, 57; Catherine, 30; Jacob Stober, 12; and property valued at $3000.*

### Christian Good

- 27 acres improved and 4 acres unimproved land valued at $1600.
- Farm machinery valued at $100.
- 1 horse, 3 milch cows, 1 other cattle, 4 swine, valued at $160.
- 30 bu. wheat, 60 bu. Indian corn, 120 bu. oats, 20 bu. Irish potatoes, 3 bu. sweet potatoes, orchard products valued at $10, 300 lbs. butter, 10 tons hay, 1 bu. clover seed.
- Animals slaughtered valued at $75.

*Federal census lists Christian, 24; Benjamin Good, 28, a farm laborer; John Good, 26, a farm laborer; Elizabeth, 29, serving; Jacob, 19, a day laborer; and property valued at $1600.*

### Samuel Good

- 75 acres improved and 45 acres unimproved land valued at $5000.
- Farm machinery valued at $200.
- 7 horses, 6 milch cows, 6 other cattle, 8 swine, valued at $500.
- 220 bu. wheat, 15 bu. rye, 300 bu. Indian corn, 300 bu. oats, 30 bu. Irish potatoes, orchard products valued at $5, 500 lbs. butter, 20 tons hay, 6 bu. clover seed, 10 bu. other grass seed.
- Animals slaughtered valued at $150.

*Federal census lists Samuel, 44; Rachael, 33; John, 21, a day laborer; Christian, 18; Jacob, 11; Maria, 23; Susanna, 16; Barbara, 9; Catherine, 8; Elizabeth, 8/12; Daniel Good, 52, a day laborer; and property valued at $5000.*

### Jacob Steffy

- 56 acres improved and 5 acres unimproved land valued at $3000.
- Farm machinery valued at $200.
- 3 horses, 4 milch cows, 11 other cattle, 4 swine, valued at $500.
- 173 bu. wheat, 11 bu. rye, 200 bu. Indian corn, 150 bu. oats, 40 bu. Irish potatoes, 2 bu. sweet potatoes, 22 bu. buckwheat, 400 lbs. butter, 12 tons hay, 1 bu. other grass seed.
- Animals slaughtered valued at $128.

*Federal census lists Jacob, 33; Susanna, 24; Baab Anna, 4; Sarah Ann, 2; Samuel L., 1; and property valued at $3000.*

### Christian Schneder

- 40 acres improved and 11 acres unimproved land valued at $2000.
- Farm machinery valued at $15.
- 1 horse, 6 milch cows, 4 other cattle, 3 swine, valued at $100.
- 80 bu. wheat, 200 bu. Indian corn, 140 bu. oats, 100 bu. Irish potatoes, 500 lbs. butter, 5 tons hay.
- Animals slaughtered valued at $60.

*Federal census lists Christian, 57; Margaret, 55; Susanna Sweigart, 17, serving.*

### Samuel Fox

- 100 acres improved and 50 acres unimproved land valued at $2000.
- Farm machinery valued at $150.
- 4 horses, 4 milch cows, 6 other cattle, 6 swine, valued at $250.

- 152 bu. wheat, 3 bu. rye, 95 bu. Indian corn, 100 bu. oats, 25 bu. Irish potatoes, 6 bu. buckwheat, 200 lbs. butter, 8 tons hay.
- Animals slaughtered valued at $45.

*Federal census lists Samuel, 58; Elizabeth, 60; William, 22, a day laborer; Barbara, 20; Magdalinea Beam, 80, a widow; and property valued at $2000.*

## Christian Burkholder

- 28 acres improved and 2 acres unimproved land valued at $2500.
- Farm machinery valued at $100.
- 2 horses, 2 milch cows, 2 other cattle, 2 swine, valued at $150.
- 50 bu. wheat, 100 bu. Indian corn, 120 bu. oats, 30 bu. Irish potatoes, 200 lbs. butter, 1 ton hay.
- Animals slaughtered valued at $45.

*Federal census lists Christian, 32, a master waggon maker; Mary, 32; John, 3; Martha, 1; Mary Weaver, 74, a widow; Adam Hornberger, 22, a waggon maker; Eliza Goshen, 13, serving; and property valued at $2500.*

## Jacob Burkhart

- 35 acres improved and 19 acres unimproved land valued at $2500.
- Farm machinery valued at $100.
- 2 horses, 4 milch cows, 1 other cattle, 5 swine, valued at $325.
- 140 bu. wheat, 12 bu. rye, 150 bu. Indian corn, 100 bu. oats, 20 bu. Irish potatoes, 22 bu. buckwheat, 200 lbs. butter, 22 tons hay.
- Animals slaughtered valued at $105.

*Federal census lists Jacob, 57; Hannah, 48; Susanna, 16; Barbara, 14; John Gaman, 4; and property valued at $2500.*

## Isaac Becker

- 55 acres improved and 9 acres unimproved land valued at $3000.
- Farm machinery valued at $52.
- 4 horses, 4 milch cows, 3 other cattle, 4 swine, valued at $300.
- 180 bu. wheat, 150 bu. Indian corn, 200 bu. oats, 50 bu. Irish potatoes, 2 bu. sweet potatoes, 200 lbs. butter, 10 tons hay, 32 lbs. honey.
- Animals slaughtered valued at $90.

*Federal census lists Isaac, 46; Sarah, 43; Caroline, 2; Susanna, 18; John Grissie, 10; and property valued at $3000.*

## Anthony Good

- 60 acres improved and 70 acres unimproved land valued at $4600.
- Farm machinery valued at $50.
- 1 horse, 3 milch cows, 10 other cattle, 8 swine, valued at $340.
- 130 bu. wheat, 200 bu. Indian corn, 235 bu. oats, 30 bu. Irish potatoes, orchard products valued at $5, 300 lbs. butter, 15 tons hay.
- Animals slaughtered valued at $105.

*Federal census lists Anthony, 48; Rachel, 38; Ellen, 8; William, 6; Caroline, 1; Jacob Good, 25, a farm laborer; and property valued at $4600.*

## John H. Good

- 75 acres improved and 10 acres unimproved land valued at $4000.
- Farm machinery valued at $100.
- 3 horses, 4 milch cows, 7 other cattle, 3 sheep, 7 swine, valued at $300.
- 150 bu. wheat, 20 bu. rye, 200 bu. Indian corn, 150 bu. oats, 9 lbs. wool, 50 bu. Irish potatoes, 6 bu. sweet potatoes, orchard products valued at $2, 200 lbs. butter, 25 tons hay.
- Animals slaughtered valued at $82.

*Federal census lists John, 46; Lavina, 36; Catherine, 15; Esther, 13; Martin, 10; Elizabeth, 7; Benjamin Haven, 18, a farm laborer; and property valued at $4000.*

## Christian Schneder

- 125 acres improved and 20 acres unimproved land valued at $7000.
- Farm machinery valued at $300.
- 4 horses, 9 milch cows, 10 other cattle, 1 sheep, 10 swine, valued at $600.
- 220 bu. wheat, 17 bu. rye, 400 bu. Indian corn, 400 bu. oats, 7 lbs. wool, 100 bu. Irish potatoes, orchard products valued at $10, 600 lbs. butter, 30 tons hay, 4 bu. clover seed, 3 bu. other grass seed.
- Animals slaughtered valued at $225.

*Federal census lists Christian, 45; Elizabeth, 45; Sarah, 20; Caroline, 18; Harrison, 15; Christian, 13; Catherine, 10; Elizabeth, 7; Mary 6; Eleanor and Clara; 3; Peter Yohn, 66, tenant (blind); Catherine Yohn, 61, tenant (blind); and property valued at $7000.*

### Christian Bleam

- 97 acres improved and 9 acres unimproved land valued at $3000.
- Farm machinery valued at $50.
- 2 horses, 4 milch cows, 6 other cattle, 2 swine, valued at $200.
- 50 bu. wheat, 20 bu. rye, 200 bu. Indian corn, 140 bu. oats, 50 bu. Irish potatoes, orchard products valued at $5, 200 lbs. butter, 20 tons hay, 2 bu. clover seed, 2 bu. other grass seed.
- Animals slaughtered valued at $75.

*Federal census lists Christian, 61; Elizabeth, 52; John Weaver, 17, a farm laborer; Savilla Peck, 16, serving; and property valued at $3000.*

### Amos Schneder

- 30 acres improved and 4 acres unimproved land valued at $2000.
- Farm machinery valued at $30.
- 2 horses, 3 milch cows, 4 other cattle, 5 sheep, 4 swine, valued at $150.
- 30 bu. wheat, 4 bu. rye, 35 bu. Indian corn, 50 bu. oats, 10 lbs. wool, 20 bu. Irish potatoes, 1 bu. sweet potatoes, 100 lbs. butter, 12 tons hay, 1 bu. clover seed, 1 bu. other grass seed, 10 lbs. flax, 2 bu. flaxseed.
- Homemade manufactures valued at $200.
- Animals slaughtered valued at $60.

*Federal census lists Amos, 53; Catherine, 52; Israel Fry, 8; and property valued at $2000.*

### Balzer Schneder

- 45 acres improved and 15 acres unimproved land valued at $2000.
- Farm machinery valued at $10.
- 2 horses, 4 milch cows, 4 other cattle, 4 swine, valued at $100.
- 50 bu. wheat, 4 bu. rye, 100 bu. Indian corn, 125 bu. oats, 50 Irish potatoes, 250 lbs. butter, 10 tons hay, 3 bu. clover seed, 1 bu. other grass seed.
- Animals slaughtered valued at $83.

*Federal census lists Balzer, 28; Elizabeth, 27; Sara, 5; Davis, 3; Mary Bausman, 59; and property valued at $2000.*

### Daniel Sensenig

- 116 acres improved and 7 acres unimproved land valued at $6000.
- Farm machinery valued at $200.
- 5 horses, 7 milch cows, 1 other cattle, 5 swine, valued at $300.
- 140 bu. wheat, 100 bu. rye, 400 bu. Indian corn, 200 bu. oats, 25 bu. Irish potatoes, 4 bu. sweet potatoes, orchard products valued at $5, 300 lbs. butter, 20 tons hay, 10 bu. other grass seed.
- Animals slaughtered valued at $150.

*Federal census lists Daniel, 48; Elizabeth, 47; Anna, 22; Jacob, 20, a farm laborer; Amos, 17, a farm laborer; Isaac, 14; Matilda, 12; John, 9; Elizabeth, 7; Daniel, 4; Anna Sensenig, 67, a widow; and property valued at $6000.*

### John Geigly

- 98 acres improved land valued at $3000.
- Farm machinery valued at $50.
- 3 horses, 3 milch cows, 2 other cattle, 6 swine, valued at $150.
- 105 bu. wheat, 250 bu. Indian corn, 200 bu. oats, 30 bu. Irish potatoes, 200 lbs. butter, 12 tons hay, 3 bu. clover seed.
- Animals slaughtered valued at $90.

*Federal census lists John, 39; Matilda, 39; Amos, 9; Magdalinea, 11; Barbara Ann, 6; Elizabeth, 4; and property valued at $3000.*

### Jacob Schneder

- 50 acres improved and 23 acres unimproved land valued at $2500.
- Farm machinery valued at $100.
- 2 horses, 4 milch cows, 5 other cattle, 8 swine, valued at $150.
- 90 bu. wheat, 20 bu. rye, 50 bu. Indian corn, 150 bu. oats, 20 bu. Irish potatoes, 200 lbs. butter, 8 tons hay.
- Animals slaughtered valued at $45.

*Federal census lists Jacob, 30; Maria, 29; Obed, 3; Mary Moore, 25, serving; and property valued at $2500.*

### Frederick Eshelman

- 80 acres improved and 12 acres unimproved land valued at $4200.
- Farm machinery valued at $50.
- 3 horses, 3 milch cows, 5 other cattle, 10 swine, valued at $500.
- 139 bu. wheat, 60 bu. rye, 150 bu. Indian corn, 150 bu. oats, 15 bu. Irish potatoes, orchard products valued at $20, 100 lbs. butter, 10 tons hay, 2 bu. clover seed, 15 lbs. flax, 1 bu. flaxseed.
- Animals slaughtered valued at $90.

*Federal census lists Frederick, 58; Elizabeth, 52; Mary, 22; Joseph, 16, a farm laborer; Moses, 7; Elizabeth, 5; and property valued at $4200.*

### Moses Messner

- 44 acres improved and 10 acres unimproved land valued at $2200.
- Farm machinery valued at $100.
- 2 horses, 3 milch cows, 3 other cattle, 5 swine, valued at $250.
- 65 bu. wheat, 30 bu. rye, 100 bu. Indian corn, 150 bu. oats, 30 bu. Irish potatoes, orchard products valued at $40, 150 lbs. butter, 10 tons hay, 2 bu. clover seed, 16 lbs. flax, 1 bu. flaxseed.
- Animals slaughtered valued at $75.

*Federal census lists Moses, 36; Elizabeth, 32; Maria, 10; Susanna, 9; Israel, 6; Moses, 4; Isaac, 1; Elizabeth Foltz, 68, a widow; and property valued at $2200.*

### Daniel Messner

- 50 acres improved and 30 acres unimproved land valued at $2500.
- Farm machinery valued at $50.
- 2 horses, 5 milch cows, 4 other cattle, 3 swine, valued at $250.
- 50 bu. wheat, 25 bu. rye, 80 bu. Indian corn, 80 bu. oats, 40 bu. Irish potatoes, 200 lbs. butter, 15 tons hay, 2 bu. clover seed, 15 lbs. flax, 1 bu. flaxseed.
- Animals slaughtered valued at $75.

*Federal census lists Daniel, 64; Catherine, 59; Elizabeth, 35; Sarah, 3; Magdalena, 24, serving; Martin Messner, 28, a day laborer; and property valued at $2500.*

### Samuel Messner

- 70 acres improved and 19 acres unimproved land valued at $5000.
- Farm machinery valued at $300.
- 3 horses, 5 milch cows, 7 other cattle, 14 swine, valued at $300.
- 216 bu. wheat, 40 bu. oats, 250 bu. Indian corn, 200 bu. oats, 20 bu. Irish potatoes, 200 lbs. butter, 20 tons hay, 1 bu. clover seed.
- Animals slaughtered valued at $75.

*Federal census lists Samuel, 49; Lydia, 52; Samuel, 11; Harriet Hornberger, 14; Jacob, 7; and property valued at $500.*

### Jacob Kile

- 21 acres improved and 9 acres unimproved land valued at $1600.
- Farm machinery valued at $30.
- 2 horses, 4 milch cows, 3 other cattle, 3 swine, valued at $150.
- 65 bu. wheat, 100 bu. Indian corn, 150 bu. oats, 10 bu. Irish potatoes, 300 lbs. butter, 7 tons hay, 2 bu. clover seed.
- Animals slaughtered valued at $52.

*Federal census lists Jacob, 48; Eliza, 35; Catherine, 11; James, 9; Anna, 5; and property valued at $1600.*

### Amos Hudison

- 64 acres improved and 6 acres unimproved land valued at $2000.
- Farm machinery valued at $150.
- 3 horses, 6 milch cows, 4 other cattle, 6 swine, valued at $200.
- 80 bu. wheat, 10 bu. rye, 100 bu. Indian corn, 150 bu. oats, 50 bu. Irish potatoes, 3 bu. sweet potatoes, orchard products valued at $5, 300 lbs. butter, 10 tons hay, 1 bu. clover seed, 1 bu. other grass seed.
- Animals slaughtered valued at $116.

*Federal census lists Amos, 51; Nancy, 52; Elias, 28, a farm laborer; Elizabeth, 26; Catherine, 17; Mary, 13; William Brown, 10; and property valued at $2000.*

# Missed on Agriculture List

### Henry Good
*Federal census lists Henry, 82; Elizabeth, 75; Elizabeth Good, 36, serving; Sarah Good, 29, serving. No property value listed. Listed next to John H. Good.*

### Emanuel Steffey
*Federal census lists Emanuel, 63; Barbara, 53; Daniel Wanerich, 12; and property valued at $400. Listed next to Jacob Steffey.*

### John Musser
*Federal census lists John, 70; Leha, 30; Susanna, 26. Property value is illegible. Listed next to Peter Musser.*

### Jacob Hoover
*Federal census lists Jacob, 67; Anna, 67; Elizabeth Stober, 35, serving; and property valued at $2100.*

### Jacob Good
*Federal census lists Jacob, 33; Judith, 27; male, 5; Anna, 8/12; Linda Ressler, 13; and property valued at $2500.*

### Jacob Mowery
*Federal census lists Jacob, 32; Sarah, 32; William, 10; Elizabeth, 8; Jacob, 6; Sarah, 1.*

### George Trostle
*Federal census lists George, 46; Caroline, 40; William, 15; Mary, 12; Samuel, 10; George, 6; Queen, 1 (female); and property valued at $1000.*

### Simon Cole
*Federal census lists Simon, 66; Mary Miller, 45, a widow; Lucy, 12; Henry, 10; Jacob, 8; Robert, 6; Benjamin, 3; Jefferson, 1; and property valued at $400.*

### Johnason Smith
*Federal census lists Johnason, 47; Elizabeth, 47; Sarah, 22; Samuel, 19; Elizabeth, 17; Lydia, 13; Catherine, 12; Jesse, 10; Johnason, 6; and property valued at $1500.*

### Joseph Garman (see Gehman)
*Federal census lists Joseph, 48; Esther, 45; Isaac, 24, a farm laborer; Henry, 16, a farm laborer; Lydia, 10; and property valued at $3500. (See Joseph Brendle, page 33.)*

### David Bixler
*Federal census lists David, 51; Susanna Bixler, 28, serving; John Lorah, 24, a day laborer; Mary Conney, 15; David Lorah, 2; John, 4/12; and property valued at $2000.*

### Richard Davis
*Federal census lists Richard, 72; Catherine, 68; Samuel, 27; Anna Fry, 19, serving; and property valued at $6000.*

# Brecknock Township—1870

## Judith Good
• 37 acres improved and 6 acres unimproved land valued at $3000.
• Farm machinery valued at $100.
• Total wages paid, $40.
• 1 horse, 4 milch cows, 3 swine, valued at $259.
• 100 bu. wheat, 4 bu. rye, 100 bu. Indian corn, 190 bu. oats, 15 bu. Irish potatoes, 2 bu. sweet potatoes, 3 gal. wine, 200 lbs. butter, 12 tons hay.
• Animals slaughtered valued at $100.
• Estimated value of farm production, $922.

*Federal census lists Judith, 37; Menno, 15, a farm laborer; Anna, 11; in same house, Anna Musselman, 60; Elizabeth, 39, a seamstress; and property valued at $3000.*

## Daniel Gehman
• 37 acres improved and 8 acres unimproved land valued at $4000.
• Farm machinery valued at $300.
• Total wages paid, $25.
• 4 horses, 6 milch cows, 6 other cattle, 4 swine, valued at $713.
• 190 bu. wheat, 20 bu. rye, 200 bu. Indian corn, 200 bu. oats, 100 bu. Irish potatoes, orchard products valued at $50, 1 gal. wine, 200 lbs. butter, 20 tons hay, 1 bu. clover seed, 1 bu. grass seed.
• Animals slaughtered valued at $170
• Estimated value of farm production, $1884.

*Federal census lists Daniel, 59; Judith, 51; John Kramer, 18, a farm laborer; Susan Trostle, 15, a domestic; and property valued at $4000.*

## George Heft
• 65 acres improved and 7 acres unimproved land valued at $4000.
• Farm machinery valued at $200.
• Total wages paid, $9.
• 2 horses, 5 milch cows, 4 other cattle, 2 sheep, 4 swine, valued at $557.
• 80 bu. wheat, 20 bu. rye, 40 bu. Indian corn, 100 bu. oats, 7 lbs. wool, 50 bu. Irish potatoes, orchard products valued at $20, 300 lbs. butter, 15 tons hay, 4 bu. clover seed, 1 bu. grass seed.
• Animals slaughtered valued at $170.
• Estimated value of farm production, $1304.

*Federal census lists George, 29; Mary, 27; Wayne, 8; Rose, 6; Samuel, 5; Charles, 2; Emma Wagner, 13; Benjamin Kessler, 67, retired farmer; Susan, 62; and property valued at $4000.*

## Amos Good
• 68 acres improved and 55 acres unimproved land valued at $6000.
• Farm machinery valued at $300.
• Total wages paid, $70.
• 3 horses, 5 milch cows, 10 other cattle, 3 sheep, 6 swine, valued at $872.
• 170 bu. wheat, 30 bu. rye, 150 bu. Indian corn, 200 bu. oats, 15 lbs. wool, 80 bu. Irish potatoes, 2 bu. sweet potatoes, orchard products valued at $25, 2 gal. wine, 300 lbs. butter, 30 tons hay, 1 bu. clover seed.
• Animals slaughtered valued at $307.
• Estimated value of farm production, $2276.

*Federal census lists Amos, 32; Lavinia, 29; Eli, 7; Savilla, 5; Elizabeth, 1; Elizabeth Steffy, 60; Elias Musser, 17, a farm laborer; and property valued at $6000.*

## John Frees
• 126 acres improved and 5 acres unimproved land valued at $8000.
• Farm machinery valued at $400.
• Total wages paid, $244.
• 6 horses, 10 milch cows, 6 other cattle, 7 swine, valued at $1175.
• 250 bu. wheat, 150 bu. Indian corn, 300 bu. oats, 150 bu. Irish potatoes, orchard products valued at $20, market gardens' produce valued at $60, 400 lbs. butter, 40 tons hay, 6 bu. clover seed, 4 bu. grass seed.
• Animals slaughtered valued at $200.
• Estimated value of farm production, $3200.

*Federal census lists John, 54; Mary, 52; John, 27, a farm laborer; Hannah, 20; Sarah, 18; Ellen, 14; Israel, 12; Emma, 10; Amanda, 8; and property valued at $8000.*

### Daniel Oberholtzer

- 100 acres improved and 28 acres unimproved land valued at $6000.
- Farm machinery valued at $400.
- Total wages paid, $187.
- 4 horses, 7 milch cows, 10 other cattle, 7 swine, valued at $850.
- 150 bu. wheat, 36 bu. rye, 200 bu. Indian corn, 350 bu. oats, 75 bu. Irish potatoes, orchard products valued at $30, 10 gal. wine, 770 lbs. butter, 30 tons hay, 1 bu. grass seed, forest products valued at $11.
- Animals slaughtered valued at $342.
- Estimated value of farm production, $2521.

*Federal census lists Daniel, 63; Sarah, 56; Susan, 29; John, 22; August Burkhart, 13; and property valued at $6000.*

### Isaac Good

- 14 acres improved and 5 acres unimproved land valued at $2000.
- Farm machinery valued at $25.
- 1 horse, 2 milch cows, 1 other cattle, 3 swine, valued at $238.
- 22 bu. wheat, 25 bu. Indian corn, 52 bu. oats, 20 bu. Irish potatoes, 1 bu. sweet potatoes, orchard products valued at $12, 5 gal. wine, 100 lbs. butter, 5 tons hay.
- Animals slaughtered valued at $80.
- Estimated value of farm production, $529.

*Federal census lists Isaac, 59; Elizabeth, 61; Magdaline Wolf, 20, a domestic; and property valued at $2000.*

### Henry Kring

- 40 acres improved and 100 acres unimproved land valued at $2500.
- Farm machinery valued at $100.
- Total wages paid, $40.
- 2 horses, 8 milch cows, 3 other cattle, 4 sheep, 4 swine, valued at $604.
- 51 bu. wheat, 60 bu. Indian corn, 60 bu. oats, 20 lbs. wool, 100 bu. Irish potatoes, orchard products valued at $20, 4 gal. wine, market gardens' produce valued at $20, 770 lbs. butter, 20 tons hay, 1 bu. clover seed, forest products valued at $5.
- Animals slaughtered valued at $129.
- Estimated value of farm production, $1495.

*Federal census lists Henry, 59; Elizabeth, 59; Elizabeth, 19; and property valued at $2500.*

### Daniel Brendel

- 85 acres improved and 2 acres unimproved land valued at $2000.
- Farm machinery valued at $50.
- Total wages paid, $18.
- 1 horse, 6 milch cows, 10 other cattle, 3 swine, valued at $380.
- 50 bu. wheat, 50 bu. Indian corn, 30 bu. Irish potatoes, orchard products valued at $10, market gardens produce valued at $15, 520 lbs. butter, 6 tons hay.
- Animals slaughtered valued at $100.
- Estimated value of farm production, $980.

*Federal census lists Daniel, 53; Susan, 45; Harrison, 17; and property valued at $2000.*

### Jacob Leinbach

- 25 acres improved land valued at $1600.
- Farm machinery valued at $100.
- Total wages paid, $12.
- 1 horse, 3 milch cows, 1 other cattle, 4 swine, valued at $260.
- 37 bu. wheat, 25 bu. Indian corn, 30 bu. oats, 60 bu. Irish potatoes, 3 bu. sweet potatoes, orchard products valued at $6, 5 gal. wine, market gardens' produce valued at $10, 150 lbs. butter, 10 tons hay, 1 bu. grass seed.
- Animals slaughtered valued at $96.
- Estimated value of farm production, $592.

*Federal census lists Jacob, 48; Susan, 44; Mary, 16; Jacob, 9; and property valued at $1600.*

### John Good

- 26 acres improved and 2 acres unimproved land valued at $1200.
- Farm machinery valued at $100.
- Total wages paid, $12.
- 1 horse, 1 milch cow, 1 other cattle, 4 swine, valued at $150.
- 45 bu. wheat, 40 bu. Indian corn, 60 bu. oats, 30 bu. Irish potatoes, orchard products valued at $25, 12 gal. wine, 100 lbs. butter, 6 tons hay.
- Animals slaughtered valued at $140.
- Estimated value of farm production, $593.

*Federal census lists John, 36; Sarah, 29; Mary A., 9; Abraham, 7; Samuel, 5; Emma, 3; and property valued at $1200.*

### Benjamin Gehman

- 57 acres improved and 1 acre unimproved land valued at $3360.
- Farm machinery valued at $190.
- 2 horses, 6 milch cows, 4 sheep, 4 swine, valued at $302.
- 25 bu. wheat, 15 bu. rye, 100 bu. Indian corn, 100 bu. oats, 10 lbs. wool, 40 bu. Irish potatoes, orchard products valued at $5, market gardens' produce valued at $10, 400 lbs. butter, 15 tons hay.
- Animals slaughtered valued at $100.
- Estimated value of farm production, $949.

*Federal census lists Benjamin, 62; Elizabeth, 57; Noe, 13 (a male); and property valued at $3360.*

### John Steffy

- 28 acres improved and 4 acres unimproved land valued at $2000.
- Farm machinery valued at $200.
- 3 horses, 3 milch cows, 1 other cattle, 4 swine, valued at $415.
- 45 bu. wheat, 18 bu. rye, 50 bu. Indian corn, 75 bu. oats, 40 bu. Irish potatoes, orchard products valued at $25, 250 lbs. butter, 6 tons hay.
- Animals slaughtered valued at $120.
- Estimated value of farm production, $890.

*Federal census lists John, 49; Susan, 45; Catherine, 25; Susan, 19; and property valued at $2000.*

### William Good

- 47 acres improved and 6 acres unimproved land valued at $4000.
- Farm machinery valued at $100.
- Total wages paid, $15.
- 2 horses, 4 milch cows, 5 other cattle, 3 swine, valued at $480.
- 75 bu. wheat, 10 bu. rye, 100 bu. Indian corn, 140 bu. oats, 30 bu. Irish potatoes, 1 bu. sweet potatoes, orchard products valued at $30, 2 gal. wine, 400 lbs. butter, 15 tons hay, 4 bu. clover seed.
- Animals slaughtered valued at $125.
- Estimated value of farm production, $1278.

*Federal census lists William, 24; Catherine, 26; William Bowman, 7; and property valued at $4000.*

### Benjamin Musser

- 60 acres improved and 30 acres unimproved land valued at $3000.
- Farm machinery valued at $200.
- Total wages paid, $60.
- 2 horses, 5 milch cows, 6 other cattle, 4 swine, valued at $510.
- 140 bu. wheat, 30 bu. rye, 150 bu. Indian corn, 150 bu. oats, 20 lbs. wool, 50 bu. Irish potatoes, orchard products valued at $100, 5 gal. wine, 500 lbs. butter, 15 tons hay.
- Animals slaughtered valued at $170.
- Estimated value of farm production, $1625.

*Federal census lists Benjamin, 63; Elizabeth, 56; Elizabeth, 23; Samuel, 21, a farm laborer; and property valued at $3000.*

### Jonas Good

- 53 acres improved and 7 acres unimproved land valued at $4000.
- Farm machinery valued at $250.
- Total wages paid, $70.
- 3 horses, 5 milch cows, 5 other cattle, 4 swine, valued at $655.
- 134 bu. wheat, 25 bu. rye, 175 bu. Indian corn, 170 bu. oats, 52 Irish potatoes, 2 bu. sweet potatoes, orchard products valued at $7, 13 gal. wine, 520 lbs. butter, 20 tons hay.
- Animals slaughtered valued at $225.
- Estimated value of farm production, $1782.

*Federal census lists Jonas, 57; Catherine, 55; John Gaiman, 14; Anna Gaiman, 13; and property valued at $4000.*

### Isaac K. Bowman

- 14 acres improved and 54 acres unimproved land valued at $3600.
- Farm machinery valued at $100.
- Total wages paid, $100.
- 2 milch cows, 4 swine, valued at $155.
- 23 bu. wheat, 50 bu. Indian corn, 103 bu. oats, 30 bu. Irish potatoes, 1 bu. sweet potatoes, orchard products valued at $50, 4 gal. wine, 300 lbs. butter, 5 tons hay.
- Animals slaughtered valued at $140.
- Estimated value of farm production, $644.

*Federal census lists Isaac, 57; Judith, 56; Mary Gehman, 49, domestic; Anna Huber, 76; and property valued at $3600.*

### Solomon Good

- 4 acres improved and 19 acres unimproved land valued at $2400.
- Farm machinery valued at $30.
- 1 horse, 1 milch cow, 2 swine, valued at $245.
- 18 bu. wheat, 20 bu. Irish potatoes, 200 lbs. butter, 1 ton hay.
- Animals slaughtered valued at $138.
- Estimated value of farm production, $482.

*Federal census lists Solomon, 39; Fanny, 31; Noe, 9; Aaron, 7; Peter, 3; John Werner, 21, a miller; Elias Gehman, 17, a farm laborer; and property valued at $2400.*

### Moses Bowman

- 53 acres improved and 15 acres unimproved land valued at $6000.
- Farm machinery valued at $200.
- Total wages paid, $40.
- 3 horses, 6 milch cows, 3 other cattle, 2 sheep, 4 swine, valued at $655.
- 115 bu. wheat, 20 bu. rye, 160 bu. Indian corn, 240 bu. oats, 3 lbs. wool, 70 bu. Irish potatoes, 3 bu. sweet potatoes, orchard products valued at $25, 10 gal. wine, 820 lbs. butter, 20 tons hay, 1 bu. clover seed, 1 bu. grass seed, forest products valued at $6.
- Animals slaughtered valued at $200.
- Estimated value of farm production, $1893.

*Federal census lists Moses, 47; Mary, 51; Gerson, 15: Anna, 11; Mary Lichty, 16, a domestic; next door, Daniel Bowman, 74, a retired farmer; Elizabeth, 71; and property valued at $6000.*

### Isaac Bowman

- 40 acres improved and 13 acres unimproved land valued at $4000.
- Farm machinery valued at $200.
- Total wages paid, $20.
- 2 horses, 3 milch cows, 5 swine, valued at $505.
- 110 bu. wheat, 10 bu. rye, 150 bu. Indian corn, 200 bu. oats, 125 bu. Irish potatoes, 6 bu. sweet potatoes, orchard products valued at $20, 8 gal. wine, 600 lbs. butter, 18 tons hay, forest products valued at $12.
- Animals slaughtered valued at $256.
- Estimated value of farm production, $1681.

*Federal census lists Isaac, 52; Lea, 55; Elias, 29; Susan, 22; Daniel, 19; Elizabeth; and property valued at $4000.*

### Jacob Musselman

- 71 acres improved and 3 acres unimproved land valued at $4000.
- Farm machinery valued at $300.
- Total wages paid, $90.
- 3 horses, 4 milch cows, 9 other cattle, 7 swine, valued at $687.
- 150 bu. wheat, 40 bu. rye, 190 bu. Indian corn, 200 bu. oats, 60 bu. Irish potatoes, 6 bu. sweet potatoes, orchard products valued at $30, 6 gal. wine, 520 lbs. butter, 18 tons hay, 3 bu. clover seed.
- Animals slaughtered valued at $242.
- Estimated value of farm production, $1924.

*Federal census lists Jacob, 33; Elizabeth, 31; Mary A., 10; Mathias, 8; Abraham, 5; Wilson, 1; and property valued at $4000.*

### H. B. Becker

- 20 acres improved and 1 acre unimproved land valued at $4500.
- Farm machinery valued at $75.
- Total wages paid, $106.
- 1 horse, 3 milch cows, 3 swine, valued at $420.
- 32 bu. wheat, 100 bu. Indian corn, 85 bu. oats, 45 bu. Irish potatoes, 1 bu. sweet potatoes, orchard products valued at $40, 3 gal. wine, 250 lbs. butter, 6 tons hay.
- Animals slaughtered valued at $207.
- Estimated value of farm production, $647.

*Federal census lists H.B., 38, a conveyeneer; Caroline, 37; John, 14; Mary, 2; Mary Trostel, 12; and property valued at $4500.*

### Adam Ressler

- 50 acres improved and 7 acres unimproved land valued at $4000.
- Farm machinery valued at $100.
- Total wages paid, $18.
- 4 horses, 4 milch cows, 8 other cattle, 4 swine, valued at $650.
- 75 bu. wheat, 20 bu. rye, 75 bu. Indian corn, 110 bu. oats, 50 bu. Irish potatoes, orchard products valued at $15, 3 gal. wine, 360 lbs. butter, 15 tons hay, 2 bu. clover seed.

- Animals valued at $177.
- Estimated value of farm production, $1443.

*Federal census lists Adam, 69; Rachel, 65; John, 25; Elizabeth, 22; Ammanda Lorah, 10; and property valued at $4000.*

### Lea Ressler
- 25 acres improved and 15 acres unimproved land valued at $1500.
- Farm machinery valued at $75.
- 1 horse, 3 milch cows, 3 other cattle, 4 swine, valued at $270.
- 40 bu. wheat, 18 bu. rye, 75 bu. Indian corn, 66 bu. oats, 30 bu. Irish potatoes, 3 bu. sweet potatoes, orchard products valued at $4, 200 lbs. butter, 7 tons hay.
- Animals slaughtered valued at $108.
- Estimated value of farm production, $726.

*Federal census lists Lea, 63; Susan, 28; Jacob, 27; and property valued at $1500.*

### Samuel Burkhart
- 80 acres improved and 49 acres unimproved land valued at $4000.
- Farm machinery valued at $40.
- 2 horses, 1 mule, 4 milch cows, 6 other cattle, 5 swine, valued at $300.
- 58 bu. wheat, 25 bu. rye, 50 bu. Indian corn, 100 bu. oats, 100 bu. Irish potatoes, 4 bu. sweet potatoes, orchard products valued at $40, 200 lbs. butter, 10 tons hay.
- Animals slaughtered valued at $126.
- Estimated value of farm production, $914.

*Federal census lists Samuel, 45; Sophia, 38; Charles, 7; Catherine, 4; Mary, 2; and property valued at $4000.*

### John Weis
- 68 acres improved and 14 acres unimproved land valued at $2500.
- Farm machinery valued at $75.
- 2 horses, 3 milch cows, 4 other cattle, 2 sheep, 6 swine, valued at $358.
- 55 bu. wheat, 18 bu. rye, 50 bu. Indian corn, 100 bu. oats, 16 lbs. wool, 25 bu. Irish potatoes, 1 bu. sweet potatoes, orchard products valued at $30, 150 lbs. butter, 14 tons hay, 2 bu. clover seed, 5 lbs. flax, forest products valued at $20, home manufactures valued at $10.
- Animals slaughtered valued at $105.
- Estimated value of farm production, $992.

*Federal census lists John, 63; Catherine, 53; Maria, 21; Isaac, 20, a farm laborer; Samuel 18, a farm laborer; Elizabeth, 57; John Shupp, 24, a farm laborer; Mary Steiner, 4; and property valued at $2500.*

### John Zinn
- 57 acres improved and 18 acres unimproved land valued at $2700.
- Farm machinery valued at $150.
- Total wages paid, $11.
- 2 horses, 4 milch cows, 5 other cattle, 2 sheep, 5 swine, valued at $436.
- 45 bu. wheat, 3 bu. rye, 50 bu. Indian corn, 85 bu. oats, 6 lbs. wool, 100 bu. Irish potatoes, 1 lb. sweet potatoes, orchard products valued at $30, 200 lbs. butter, 10 tons hay, 1 bu. clover seed.
- Animals slaughtered valued at $126.
- Estimated value of farm production, $1009.

*Federal census lists John, 45; Lydia, 37; Susan, 18; Fyanna, 14; Jacob, 9; Catherine, 3; Sophia Shup, 57; Susan, 19, a domestic; and property valued at $2500.*

### David Good
- 98 acres improved and 19 acres unimproved land valued at $5000.
- Farm machinery valued at $300.
- Total wages paid, $180.
- 2 horses, 5 milch cows, 8 other cattle, 4 sheep, 4 swine, valued at $592.
- 100 bu. wheat, 6 bu. rye, 280 bu. Indian corn, 220 bu. oats, 30 lbs. wool, 40 bu. Irish potatoes, 4 bu. sweet potatoes, orchard products valued at $25, 225 lbs. butter, 30 tons hay, 1 bu. clover seed, 1 bu. grass seed, forest products valued at $125.
- Animals slaughtered valued at $225.
- Estimated value of farm production, $2009.

*Federal census lists David, 39; Sarah, 32; Leman Shirk, 9; and property valued at $5000.*

### Esaias Burkholder
- 27 acres improved and 7 acres unimproved land valued at $2000.
- Farm machinery valued at $75.
- 1 horse, 2 milch cows, 1 other cattle, 2 swine, valued at $112.

- 40 bu. wheat, 80 bu. Indian corn, 80 bu. oats, 30 bu. Irish potatoes, 2 bu. sweet potatoes, orchard products valued at $25, 5 gal. wine, 150 lbs. butter, 4 tons hay, 7 lbs. flax, forest products valued at $16.
- Animals slaughtered valued at $82.
- Estimated value of farm production, $532.

*Federal census lists Esaias, 39; Sarah, 33; Caroline, 13; Franklin, 11; Wm. 7; next door, Ulrich, 77, and Nancy, 72, retired farmer; and property valued at $2000.*

### Abraham Miller

- 19 acres improved and 1 acre unimproved land valued at $2000.
- Farm machinery valued at $125.
- Total wages paid, $45.
- 1 horse, 3 milch cows, 1 other cattle, 2 swine, valued at $200.
- 29 bu. wheat, 9 bu. rye, 50 bu. Indian corn, 50 bu. oats, 30 bu. Irish potatoes, 2 bu. sweet potatoes, orchard products valued at $12, 200 lbs. butter, 5 tons hay.
- Animals slaughtered valued at $144.
- Estimated value of farm production, $598.

*Federal census lists Abraham, 26; Margaret, 24; Emma, 6; Catherine, 4; and property valued at $2000.*

### Joel Eshelman

- 30 acres improved and 170 acres unimproved land valued at $3000.
- Farm machinery valued at $100.
- 1 mule, 2 milch cows, 7 other cattle, 2 swine, valued at $235.
- 15 bu. rye, 60 bu. Irish potatoes, orchard products valued at $10, 200 lbs. butter, 8 tons hay, forest products valued at $300.
- Animals slaughtered valued at $70.
- Estimated value of farm production, $830.

*Federal census lists Joel, 55; Hannah, 52; Abraham, 7; Hannah, 4; Barbara, 6; Elizabeth, 4; Aaron, 6/12; and property valued at $3000.*

### Christ. Burkhard

- 40 acres improved and 10 acres unimproved land valued at $4000.
- Farm machinery valued at $25.
- 1 horse, 3 milch cows, 1 swine, valued at $171.
- 40 bu. wheat, 20 bu. rye, 100 bu. Indian corn, 100 bu. oats, 10 bu. Irish potatoes, 2 bu. sweet potatoes, 100 lbs. butter, 4 tons hay, 1 bu. clover seed, forest products valued at $100.
- Animals slaughtered valued at $200.
- Estimated value of farm production, $743.

*Federal census lists Christ., 33; Mary, 31; Ezra, 8; Harvy, 6; Catherine, 4; John, 2; Elizabeth, 8/12; and property valued at $4000.*

### Richard Beam

- 42 acres improved and 6 acres unimproved land valued at $3000.
- Farm machinery valued at $300.
- Total wages paid, $30.
- 2 horses, 4 milch cows, 4 other cattle, 4 swine, valued at $500.
- 100 bu. wheat, 10 bu. rye, 100 bu. Indian corn, 150 bu. oats, 20 bu. Irish potatoes, 10 bu. sweet potatoes, orchard products valued at $30, 400 lbs. butter, 15 tons hay, 4 bu. clover seed, 2 bu. grass seed.
- Animals slaughtered valued at $229.
- Estimated value of farm production, $1448.

*Federal census lists Richard, 48; Susan, 19; Sarah, 17; Mary, 14; and property valued at $3000.*

### Christian and Benjamin Good

- 35 ares improved and 3 acres unimproved land valued at $3000.
- Farm machinery valued at $200.
- Total wages paid, $15.
- 1 horse, 4 milch cows, 3 swine, valued at $386.
- 80 bu. wheat, 100 bu. Indian corn, 100 bu. oats, 20 bu. Irish potatoes, 3 bu. sweet potatoes, orchard products valued at $20, 300 lbs. butter, 15 tons hay.
- Animals slaughtered valued at $190.
- Estimated value of farm production, $1661.

*Federal census lists Christian, 34; Ben, 38; Elizabeth, 39; Henry Trostle, 10; Emmaline Coleman, 10; and property valued at $3000.*

## Solomon Ott

- 58 acres improved and 3 acres unimproved land valued at $4000.
- Farm machinery valued at $100.
- Total wages paid, $100.
- 2 horses, 4 milch cows, 3 other cattle, 3 swine, valued at $510.
- 140 bu. wheat, 40 bu. rye, 250 bu. Indian corn, 250 bu. oats, 30 bu. Irish potatoes, orchard products valued at $50, 250 lbs. butter, 25 tons hay, 1 bu. clover seed.
- Animals slaughtered valued at $100.
- Estimated value of farm production, $1686.

*Federal census lists Solomon, 39; Lea, 39; Mary, 6; Elisabeth, 2; August Hoshaur, 15; and property valued at $4000.*

## Jacob Burkhart

- 17 acres improved land valued at $2000.
- Farm machinery valued at $200.
- Total wages paid, $30.
- 2 horses, 1 milch cow, 1 other cattle, 1 swine, valued at $206.
- 35 bu. wheat, 15 bu. rye, 35 bu. Indian corn, 80 bu. oats, 20 bu. Irish potatoes, orchard products valued at $5, 100 lbs. butter, 4 tons hay, 2 bu. clover seed.
- Animals slaughtered valued at $123.
- Estimated value of farm production, $576.

*Federal census lists Jacob, 43; Rebecca, 40, Lewis, 13; Sarah, 12; Henry, 6; Lydia, 1; and property valued at $2000. Census also lists Jacob, 27; Eliza, 29; Eli, 2; Henry Stover, 11; and property valued at $2000.*

## Elizabeth Schnader

- 123 acres improved and 12 acres unimproved land valued at $9000.
- Farm machinery valued at $150.
- Total wages paid, $200.
- 4 horses, 6 milch cows, 4 other cattle, 8 swine, valued at $1231.
- 300 bu. wheat, 45 bu. rye, 400 bu. Indian corn, 760 bu. oats, 75 bu. Irish potatoes, 6 bu. sweet potatoes, orchard products valued at $60, 500 lbs. butter, 50 tons hay, 7 bu. clover seed.
- Animals slaughtered valued at $354.
- Estimated value of farm production, $3787.

*Federal census lists Elizabeth, 58; Sarah, 29; Christian, 23, a farmer; Elizabeth, 18; Mary, 16; Clara, 13; William Beam, 21, a farm laborer; Catherine Yohn, 80; and property valued at $9000.*

## Anthony Good

- 65 acres improved and 58 acres unimproved land valued at $4500.
- Farm machinery valued at $150.
- Total wages paid, $100.
- 3 horses, 5 milch cows, 5 other cattle, 6 sheep, 6 swine, valued at $523.
- 80 bu. wheat, 150 bu. Indian corn, 200 bu. oats, 40 bu. Irish potatoes, 3 bu. sweet potatoes, orchard products valued at $50, 250 lbs. butter, 20 tons hay, 4 bu. clover seed, 1/2 bu. grass seed, forest products valued at $30.
- Animals slaughtered valued at $189.
- Estimated value of farm production, $1535.

*Federal census lists Anthony, 59; Rachel, 47; Ellen, 18; William, 16; Caroline, 11; Amos, 8; Edward, 6; Jacob, 32, a farm laborer; and property valued at $4500.*

## Samuel Reifsnyder

- 62 acres improved and 15 acres unimproved land valued at $4500.
- Farm machinery valued at $150.
- Total wages paid, $125.
- 1 horse, 5 milch cows, 1 other cattle, 8 swine, valued at $340.
- 208 bu. wheat, 2 bu. rye, 225 bu. Indian corn, 300 bu. oats, 30 bu. Irish potatoes, orchard products valued at $30, 250 lbs. butter, 20 tons hay, 9 bu. clover seed, 1/2 bu. grass seed.
- Animals slaughtered valued at $684.
- Estimated value of farm production, $2172.

*Federal census lists Samuel, 33; Caroline, 29; Charles, 7; Sallie, 4; Isaac, 2; William, 2/12; Eliza Laush, 19, a domestic; Isaac Becker, 56, a retired farmer; Sallie Becker, 53; and property valued at $4500.*

## Christian Schnader

- 43 acres improved and 1 acres unimproved land valued at $3000.
- Farm machinery valued at $100.
- Total wages paid, $30.
- 2 horses, 4 milch cows, 1 other cattle, 2 swine, valued at $320.

- 70 bu. wheat, 10 bu. rye, 112 bu. Indian corn, 75 bu. oats, 75 bu. Irish potatoes, 5 bu. sweet potatoes, orchard products valued at $5, 200 lbs. butter, 6 tons hay, 1 bu. grass seed.
- Animals slaughtered valued at $134.
- Estimated value of farm production, $881.

*Federal census lists David Shaeffer, 37; Anna, 35; Samuel, 13; William, 3; Christian Schnader, 68; and property valued at $3000.*

## Jacob Steffy

- 55 acres improved and 21 acres unimproved land valued at $5400.
- Farm machinery valued at $300.
- Total wages paid, $38.
- 3 horses, 5 milch cows, 10 other cattle, 10 sheep, 8 swine, valued at $1153.
- 200 bu. wheat, 10 bu. rye, 200 bu. Indian corn, 200 bu. oats, 35 lbs. wool, 100 bu. Irish potatoes, 10 bu. sweet potatoes, orchard products valued at $15, 9 gal. wine, market gardens' produce valued at $50, 150 lbs. butter, 25 tons hay, 7 bu. clover seed, 8 bu. grass seed.
- Forest products valued at $50.
- Animals slaughtered valued at $274.
- Estimated value of farm production, $2690.

*Federal census lists Jacob, 45; Susan, 32; Barbara, 15; Mary Ellen, 13; Samuel, 11; Anna, 9; Cyrus, 7; Emanuel, 5; Lydia, 3; Jacob, 2; and property valued at $5400.*

## Isaac Shup

- 35 acres improved and 25 acres unimproved land valued at $3500.
- Farm machinery valued at $100.
- Total wages paid, $25.
- 2 horses, 3 milch cows, 11 other cattle, 4 swine, valued at $475.
- 110 bu. wheat, 100 bu. Indian corn, 150 bu. oats, 50 bu. Irish potatoes, 2 bu. sweet potatoes, orchard products valued at $15, 5 gal. wine, market gardens' produce valued at $100, 500 lbs. butter, 28 tons hay, 3 bu. clover seed, forest products valued at $12.
- Animals slaughtered valued at $171.
- Estimated value of farm production, $1686.

*Federal census lists Isaac, 33; Barbara, 27; Nathan, 6; Barton, 5; Emma, 3; Henry, 1; and property valued at $3500.*

## Benjamin Witwer

- 28 acres improved land valued at $1000.
- Farm machinery valued at $150.
- 1 horse, 2 milch cows, 3 swine, valued at $135.
- 45 bu. wheat, 2 bu. rye, 100 bu. Indian corn, 65 bu. oats, 40 bu. Irish potatoes, 1 bu. sweet potatoes, orchard products valued at $5, market gardens' produce valued at $100, 200 lbs. butter, 6 tons hay.
- Animals slaughtered valued at $164.
- Estimated value of farm production, $745.

*Federal census lists Benjamin, 29; Maria, 32; James, 8; Benjamin, 7; John, 4; Sarah, 2; Margaret, 2; and property valued at $1000.*

## Aaron Coleman

- 22 acres improved and 2 acres unimproved land valued at $1400.
- Farm machinery valued at $75.
- 1 horse, 2 milch cows, 2 other cattle, 3 swine, valued at $305.
- 35 bu. wheat, 60 bu. Indian corn, 40 bu. oats, 50 bu. Irish potatoes, 8 bu. sweet potatoes, orchard products valued at $8, 6 gal. wine, market gardens' produce valued at $40, 200 lbs. butter, 8 tons hay.
- Animals slaughtered valued at $138.
- Estimated value of farm production, $821.

*Federal census lists Aaron, 43; Barbara, 38; John, 18; Fanny, 8; and property valued at $1400.*

## Isaac Kern

- 94 acres improved and 6 acres unimproved land valued at $3000.
- Farm machinery valued at $100.
- 4 horses, 5 milch cows, 5 other cattle, 3 sheep, 7 swine, valued at $674.
- 200 bu. wheat, 200 bu. Indian corn, 200 bu. oats, 75 bu. Irish potatoes, 1 bu. sweet potatoes, orchard products valued at $10, 400 lbs. butter, 20 tons hay.
- Animals slaughtered valued at $174.
- Estimated value of farm production, $1836.

*Federal census lists Isaac, 58; Elizabeth, 24; Martin, 23, a farm laborer; John, 21, a farm laborer; Mary, 18; Susan, 16; Catherine, 14; Milton, 12; Charles Cleck, 4; and property valued at $3000.*

### John Steffy
- 50 acres improved and 40 acres unimproved land valued at $4500.
- Farm machinery valued at $400.
- Total wages paid, $140.
- 3 horses, 4 milch cows, 12 other cattle, 4 swine, valued at $800.
- 100 bu. wheat, 30 bu. rye, 200 bu. Indian corn, 180 bu. oats, 40 bu. Irish potatoes, orchard products valued at $20, 500 lbs. butter, 20 tons hay, 5 bu. clover seed, 3 bu. grass seed, 10 lbs. honey.
- Animals slaughtered valued at $162.
- Estimated value of farm production, $1911.

*Federal census lists John, 48; Louisa, 36; Margaret, 27; Henry Foltz, 12; and property valued at $4500.*

### John Good
- 73 acres improved and 5 acres unimproved land valued at $5000.
- Farm machinery valued at $150.
- 3 horses, 4 milch cows, 8 other cattle, 3 swine, valued at $545.
- 75 bu. wheat, 10 bu. rye, 125 bu. Indian corn, 100 bu. oats, 75 bu. Irish potatoes, 5 bu. sweet potatoes, orchard products valued at $15, 400 lbs. butter, 15 tons hay, 3 bu. clover seed, 3 bu. grass seed.
- Animals slaughtered valued at $134.
- Estimated value of farm production, $1366.

*Federal census lists John, 55; Lydia, 50; Allison, 21, a farm laborer; Aaron, 18, a farm laborer; Maria, 15; and property valued at $5000.*

### Christian Stauffer
- 16 acres improved land valued at $4000.
- Farm machinery valued at $150.
- Total wages paid, $30.
- 1 horse, 2 milch cows, 3 swine, valued at $329.
- 60 bu. wheat, 100 bu. Indian corn, 150 bu. oats, 30 bu. Irish potatoes, orchard products valued at $5, 150 lbs. butter, 6 tons hay.
- Animals slaughtered valued at $85.
- Estimated value of farm production, $802.

*Federal census lists Christian, 47; Julia, 57; John, 18, a butcher; Israel, 15; and property valued at $4000.*

### Isaac Burkholder
- 18 acres improved and 10 acres unimproved land valued at $2000.
- Farm machinery valued at $75.
- Total wages paid, $10.
- 1 horse, 2 milch cows, 1 other cattle, 3 swine, valued at $300.
- 30 bu. wheat, 18 bu. rye, 50 bu. Indian corn, 70 bu. oats, 40 bu. Irish potatoes, orchard products valued at $25, 8 gal. wine, orchard products valued at $30, 200 lbs. butter, 5 tons hay, 1 bu. clover seed, forest products valued at $17.
- Animals slaughtered valued at $117.
- Estimated value of farm production, $793.

*Federal census lists Isaac, 45; Catherine, 57; Maria, 20; Anna Blankenbiller, 10; and property valued at $2000.*

### Daniel Musser
- 57 acres improved and 6 acres unimproved land valued at $4500.
- Farm machinery valued at $60.
- Total wages paid, $30.
- 1 horse, 3 milch cows, 3 swine, valued at $260.
- 190 bu. wheat, 14 bu. rye, 150 bu. Indian corn, 200 bu. oats, 30 bu. Irish potatoes, 1 bu. sweet potatoes, orchard products valued at $20, 400 lbs. butter, 15 tons hay, 1 bu. clover seed.
- Animals slaughtered valued at $147.
- Estimated value of farm production, $1273.

*Federal census lists Daniel, 29; Maria, 26; Theodore, 5; Ida, 3; Isaac, 1; and property valued at $4500.*

### David Musser
- 79 acres improved and 16 acres unimproved land valued at $6000.
- Farm machinery valued at $200.
- Total wages paid, $125.
- 5 horses, 5 milch cows, 7 other cattle, 6 swine, valued at $655.
- 180 bu. wheat, 5 bu. rye, 200 bu. Indian corn, 300 bu. oats, 40 bu. Irish potatoes, 6 bu. sweet potatoes, orchard produce valued at $50, 400 lbs. butter, 20 tons hay, 4 bu. clover seed.
- Animals slaughtered valued at $209.
- Estimated value of farm production, $1905.

*Federal census lists David, 41; Hannah, 32; Samuel, 13; Sarah, 11; Henry, 8; Martin, 6; Nancy, 59, a domestic; and property valued at $6000.*

### George Lorah
- 113 acres improved and 68 acres unimproved land valued at $8000.
- Farm machinery valued at $300.
- Total wages paid, $250.
- 6 horses, 8 milch cows, 14 other cattle, 6 swine, valued at $1290.
- 300 bu. wheat, 30 bu. rye, 400 bu. Indian corn, 500 bu. oats, 40 bu. Irish potatoes, orchard products valued at $10, 400 lbs. butter, 20 tons hay, 4 bu. clover seed, 2 bu. grass seed.
- Animals slaughtered valued at $194.
- Estimated value of farm production, $2981.

*Federal census lists George, 24; Sarah, 27; Shiffer Elizabeth [Elizabeth Shiffer?], 16, a domestic; Peter Weaver, 65, a farm laborer; and property valued at $8000.*

### Samuel Schlabach
- 20 acres improved land valued at $800.
- Total wages paid, $30.
- 3 milch cows, 1 other cattle, 2 swine, valued at $160.
- 32 bu. wheat, 6 bu. rye, 45 bu. Indian corn, 20 bu. oats, 25 bu. Irish potatoes, orchard products valued at $5, 175 lbs. butter, 3 tons hay.
- Animals slaughtered valued at $95.
- Estimated value of farm production, $457.

*Federal census lists Samuel, 43; Julia, 36; Reuben, 16; Daniel, 8; and property valued at $800.*

### John Sollenberger
- 103 acres improved and 37 acres unimproved land valued at $8000.
- Farm machinery valued at $550.
- Total wages paid, $175.
- 4 horses, 8 milch cows, 22 other cattle, 1 sheep, 6 swine, valued at $1325.
- 335 bu. wheat, 10 bu. rye, 250 bu. Indian corn, 350 bu. oats, 100 bu. Irish potatoes, 4 bu. sweet potatoes, orchard products valued at $20, 4 gal. wine, 1200 lbs. butter, 78 tons hay, 2 bu. clover seed, 2 bu. grass seed, 15 lbs. flax, 1 bu. flaxseed, forest products valued at $14.
- Animals slaughtered valued at $329.
- Estimated value of farm production, $4098.

*Federal census lists John, 26; Mary, 27; Alice, 1; Rudolph Steffy, 21; Henry Steiner, 69, a farm laborer; and property valued at $8000.*

### Peter Coldren
- 24 acres improved and 6 acres unimproved land valued at $1500.
- Farm machinery valued at $50.
- 1 horse, 3 milch cows, 1 other cattle, 3 swine, valued at $150.
- 45 bu. wheat, 30 bu. Indian corn, 70 bu. oats, 25 bu. Irish potatoes, 1 gal. wine, 100 lbs. butter, 5 tons hay, 6 lbs. flax.
- Animals slaughtered valued at $85.
- Estimated value of farm production, $470.

*Federal census lists Peter, 67; Catherine, 63; Urias Steiner, 19; and property valued at $1500.*

### Benjamin Good
- 12 acres improved and 3 acres unimproved land valued at $1200.
- Farm machinery valued at $50.
- Total wages paid, $5.
- 1 horse, 1 milch cow, 2 other cattle, 3 swine, valued at $230.
- 21 bu. wheat, 50 bu. Indian corn, 25 bu. Irish potatoes, 100 lbs. butter, 5 tons hay, forest products valued at $10.
- Animals slaughtered valued at $126.
- Estimated value of farm production, $549.

*Federal census lists Benjamin, 39; Sarah, 48; Elizabeth, 14; Leman, 12; James, 8; Isaac, 42, a stone mason; and property valued at $1200.*

### John Coldren
- 59 acres improved and 9 acres unimproved land valued at $1600.
- Farm machinery valued at $100.
- Total wages paid, $7.
- 4 horses, 4 milch cows, 4 other cattle, 1 sheep, 6 swine, valued at $523.
- 150 bu. wheat, 10 bu. rye, 100 bu. Indian corn, 150 bu. oats, 30 bu. Irish potatoes, orchard products valued at $5, 5 gal. wine, 400 lbs. butter, 20 tons hay, 1 bu. grass seed.
- Animals slaughtered valued at $222.

• Estimated value of farm production, $1552.

*Federal census lists John, 66; Barbara, 67; Samuel, 23, a miller; Solomon, 20, a farm laborer; Jacob, 16, a farm laborer; and property valued at $1600.*

## George Trostel

• 56 acres improved land valued at $2000.
• Farm machinery valued at $100.
• 3 horses, 3 milch cows, 2 other cattle, 1 sheep, 6 swine, valued at $373.
• 70 bu. wheat, 5 bu. rye, 40 bu. Indian corn, 75 bu. oats, 36 bu. Irish potatoes, 1 bu. sweet potatoes, orchard products valued at $10, 150 lbs. butter, 10 tons hay.
• Animals slaughtered valued at $127.
• Estimated value of farm production, $885.

*Federal census lists George, 58; Caroline, 50; Samuel, 21; Wayne, 11; Emmaline, 7; Henry Lewis, 6; and property valued at $2000.*

## Samuel Steiner

• 10 acres improved and 3 acres unimproved land valued at $1500.
• Farm machinery valued at $30.
• Total wages paid, $12.
• 1 horse, 1 milch cow, 1 other cattle, 3 swine, valued at $185.
• 25 bu. wheat, 25 bu. Indian corn, 40 bu. Irish potatoes, 2 gal. wine, 75 lbs. butter, 3 tons hay.
• Animals slaughtered valued at $65.
• Estimated value of farm production, $392.

*Federal census lists Samuel, 37, a stone mason; Elizabeth, 28; Margaret, 3; David, 8/12; Susannah, 57; and property valued at $1500.*

## Isaac Coldren

• 68 acres improved and 15 acres unimproved land valued at $3300.
• Farm machinery valued at $150.
• Total wages paid, $150.
• 3 horses, 4 milch cows, 4 other cattle, 3 swine, valued at $415.
• 142 bu. wheat, 6 bu. rye, 100 bu. Indian corn, 100 bu. oats, 30 bu. Irish potatoes, 5 bu. sweet potatoes, 2 gal. wine, 375 lbs. butter, 15 tons hay, 2 bu. clover seed, 2 bu. grass seed.
• Animals slaughtered valued at $120.
• Estimated value of farm production, $1233.

*Federal census lists Isaac, 42; Ellen, 31; Mary, 5; Elizabeth Ruth, 15, a domestic; and property valued at $3300.*

## Otto Roland

• 27 acres improved and 7 acres unimproved land valued at $1000.
• Farm machinery valued at $75.
• Total wages paid, $4.
• 1 horse, 2 milch cows, 2 other cattle, 3 swine, valued at $213.
• 31 bu. wheat, 4 bu. rye, 35 bu. Indian corn, 22 bu. oats, 25 bu. Irish potatoes, orchard products valued at $5, 175 lbs. butter, 3 tons hay.
• Animals slaughtered valued at $72.
• Estimated value of farm production, $480.

*Federal census lists Otto, born in Prussia, 35, a tailor; Lydia, 45; Joseph, 9; August, 5; Otto, 7; Emma, 1; and property valued at $1000.*

## Jacob Gross

• 60 acres improved and 20 acres unimproved land valued at $3000.
• Farm machinery valued at $100.
• 2 horses, 3 milch cows, 4 other cattle, 6 sheep, 5 swine, valued at $500.
• 165 bu. wheat, 200 bu. Indian corn, 150 bu. oats, 30 bu. Irish potatoes, orchard products valued at $10, 500 lbs. butter, 10 tons hay.
• Animals slaughtered valued at $150.
• Estimated value of farm production, $1397.

*Federal census lists Jacob, 38; Willhelmena, 42; Charles, 14; Willhelmena, 7; Frederic, 6; Laurietta, 5/12; and property valued at $3000. Mother and Father born in Wurtemburg; children born in New Jersey.*

## Isaac Roth

• 55 acres improved and 75 acres unimproved land valued at $4000.
• Farm machinery valued at $100.
• Total wages paid, $30.
• 3 horses, 8 milch cows, 6 other cattle, 6 sheep, 4 swine, valued at $690.

• 125 bu. wheat, 5 bu. rye, 100 bu. Indian corn, 100 bu. oats, 75 bu. Irish potatoes, orchard products valued at $30, 520 lbs. butter, 25 tons hay, 3 bu. clover seed, 10 lbs. flax, 1 bu. flaxseed.
• Animals slaughtered valued at $336.
• Estimated value of farm production, $1934.

*Federal census lists Isaac, 53; Fyanna, 34; Franklin, 24; Elizabeth, 14; Lydia, 13; Emmaline, 9; David, 8; Adaline, 6; Jacob, 4; Isaac, 2; and property valued at $4000.*

## Elias Kramer

• 120 acres improved and 8 acres unimproved land valued at $3000.
• Farm machinery valued at $56.
• Total wages paid, $10.
• 4 milch cows, 8 other cattle, 7 swine, valued at $150.
• 60 bu. wheat,10 bu. rye, 100 bu. Indian corn, 125 bu. oats, 50 bu. Irish potatoes, orchard products valued at $10, 400 lbs. butter, 12 tons hay, 4 bu. clover seed.
• Animals slaughtered valued at $188.
• Estimated value of farm production, $923.

*Federal census lists Elias, 40, a collier; Lea, 29; Mary, 11; Lea, 9; Christiana, 6; Hannah, 4; and property valued at $6480.*

## John Schnader

• 30 acres improved and 4 acres unimproved land valued at $1500.
• Farm machinery valued at $40.
• 1 horse, 2 milch cows, 1 other cattle, 4 swine, valued at $100.
• 55 bu. wheat, 50 bu. Indian corn, 55 bu. oats, 30 bu. Irish potatoes, orchard products valued at $20, 200 lbs, butter, 5 tons hay, 30 lbs. flax, 2 bu. flaxseeed.
• Animals slaughtered valued at $168.
• Estimated value of farm production, $581.

*Federal census lists John, 38, a stone mason; Polly, 37; Maria, 14; John, 10; Benjamin Good, 67, a cooper; Richard Good, 29, a stone mason; and property valued at $1500.*

## Samuel Kieffer

• 73 acres improved and 20 acres unimproved land valued at $3500.
• Farm machinery valued at $50.
• Total wages paid, $125.
• 3 horses, 5 milch cows, 5 other cattle, 4 swine, valued at $371.
• 90 bu. wheat, 8 bu. rye, 50 bu. Indian corn, 95 bu. oats, 50 bu. Irish potatoes, orchard products valued at $20, 300 lbs. butter, 10 tons hay, 1 bu. clover seed.
• Animals slaughtered valued at $148.
• Estimated value of farm production, $1015.

*Federal census lists Samuel, 59; Susan, 58; Elizabeth, 17; Jeremiah Bicher, 22; and property valued at $3500.*

## Elizabeth Kieffer

• 27 acres improved and 4 acres unimproved land valued at $750.
• Total wages paid, $60.
• 3 milch cows, 1 other cattle, 3 swine, valued at $110.
• 25 bu. wheat, 30 bu. Indian corn, 30 bu. oats, 25 bu. Irish potatoes, orchard products valued at $10, 175 lbs. butter, 4 tons hay.
• Animals slaughtered valued at $84.
• Estimated value of farm production, $392.

*Federal census lists Elizabeth, 66; Eliza, 25, Isaac, 66, a shoemaker; and property valued at $750.*

## Isaac Shup

• 45 acres improved and 30 acres unimproved land valued at $2000.
• Farm machinery valued at $75.
• Total wages paid, $38.
• 2 horses, 4 milch cows, 1 other cattle, 4 sheep, 5 swine, valued at $287.
• 40 bu. wheat, 6 bu. rye, 40 bu. Indian corn, 100 bu. oats, 5 lbs. wool, 50 bu. Irish potatoes, orchard products valued at $25, 250 lbs. butter, 10 tons hay.
• Animals slaughtered valued at $196.
• Estimated value of farm production, $891.

*Federal census lists Isaac, 60; Harriet, 59; Reuben, 19; Lydia, 19; Kitty, 11; and property valued at $2000.*

## Israel Gehman

• 102 acres improved and 6 acres unimproved land valued at $7000.
• Farm machinery valued at $100.
• Total wages paid, $180.

- 4 horses, 7 milch cows, 9 other cattle, 4 swine, valued at $495.
- 80 bu. wheat, 10 bu. rye, 100 bu. Indian corn, 100 bu. oats, 40 bu. Irish potatoes, orchard products valued at $75, 400 lbs. butter, 25 tons hay.
- Animals slaughtered valued at $114.
- Estimated value of farm production, $1332.

*Federal census lists Israel, 25; Susan, 27; Benjamin, 5; Martin, 3; Elizabeth, 2; David, 1; John, 17; and property valued at $7000.*

## Jacob Bicher

- 42 acres improved and 4 acres unimproved land valued at $2000.
- Farm machinery valued at $100.
- Total wages paid, $30.
- 2 horses, 3 milch cows, 4 other cattle, 6 swine, valued at $410.
- 115 bu. wheat, 45 bu. rye, 150 bu. Indian corn, 75 bu. oats, 50 bu. Irish potatoes, 200 lbs. butter, 7 tons hay, 3 bu. clover seed, 1 bu. grass seed.
- Animals slaughtered valued at $150.
- Estimated value of farm production, $1123.

*Federal census lists Jacob, 52; Sarah, 50; Rebecca, 13; John, 14; Mary, 8; Benjamin, 6; Jacob, 2; Laura Grist, 2; and property valued at $2000.*

## Amos Hutchinson

- 15 acres improved and 7 acres unimproved land valued at $3000.
- Farm machinery valued at $50.
- Total wages paid, $10.
- 1 horse, 1 milch cow, 2 swine, valued at $250.
- 30 bu. wheat, 50 bu. Indian corn, 100 bu. oats, 30 bu. Irish potatoes, 2 bu. sweet potatoes, orchard products valued at $20, 1 gal. wine, 150 lbs. butter, 3 tons hay, forest products valued at $10.
- Animals slaughtered valued at $112.
- Estimated value of farm production, $622.

*Federal census lists Amos, 61; Anna, 61; Elizabeth, 27; and property valued at $3000.*

## Isaac Kessler

- 70 acres improved and 14 acres unimproved land valued at $4000.
- Farm machinery valued at $175.
- Total wages paid, $60.
- 4 horses, 7 milch cows, 3 other cattle, 4 sheep, 5 swine, valued at $562.
- 70 bu. wheat, 20 bu. rye, 50 bu. Indian corn, 50 bu. oats, 25 lbs. wool, 25 bu. Irish potatoes, 3 bu. sweet potatoes, orchard products valued at $10, 3 gal. wine, 600 lbs. butter, 15 tons hay, 5 bu. grass seed.
- Animals slaughtered valued at $172.
- Estimated value of farm production, $1357.

*Federal census lists Isaac, 57; Catherine, 57; Margaret, 25; Daniel, 21; Isaac, 15; Elizabeth Stover, 14; and property valued at $4000.*

## Levi Steffy

- 35 acres improved and 4 acres unimproved land valued at $2700.
- Farm machinery valued at $150.
- Total wages paid, $10.
- 2 horses, 3 milch cows, 3 other cattle, 3 swine, valued at $422.
- 50 bu. wheat, 10 bu. rye, 40 bu. Indian corn, 75 bu. oats, 100 lbs. Irish potatoes, orchard products valued at $10, 1 gal. wine, market gardens' produce valued at $5, 250 lbs. butter, 8 tons hay, 2 bu. clover seed, 1 bu. grass seed
- Animals slaughtered valued at $127,
- Estimated value of farm production, $967.

*Federal census lists Levi, 58; Rachel, 54; William, 24; and property valued at $2700.*

## Christian Good

- 68 acres improved and 4 acres unimproved land valued at $3500.
- Farm machinery valued at $100.
- Total wages paid, $12.
- 2 horses, 3 milch cows, 7 other cattle, 3 swine, valued at $422.
- 40 bu. wheat, 19 bu. rye, 120 bu. Indian corn, 160 bu. oats, 80 bu. Irish potatoes, 1 bu. sweet potatoes, orchard products valued at $20, 2 gal. wine, 350 lbs. butter, 10 tons hay, 1 bu. clover seed, 10 lbs. flax, 1 bu. flaxseed, forest products valued at $24.
- Animals slaughtered valued at $133.
- Estimated value of farm production, $1141.

*Federal census lists Christian, 70; Susan, 55; John, 20; and property valued at $3500.*

### John Brendel
- 50 acres improved and 50 acres unimproved land valued at $3000.
- Farm machinery valued at $100.
- Total wages paid, $12.
- 2 horses, 4 milch cows, 2 swine, valued at $310.
- 50 bu. wheat, 25 bu. rye, 150 bu. Indian corn, 200 bu. oats, 50 bu. Irish potatoes, 150 lbs. butter, 6 tons hay.
- Animals slaughtered valued at $142.
- Estimated value of farm production, $924.

*Federal census lists John, 30; Mary 29; John, 7; Susan, 5; Hiram, 2; Martha 5/12; Beneville Redcay, 14; and property valued at $3000.*

### Joseph Zerbe
- 28 acres improved and 28 acres unimproved land valued at $2000.
- Farm machinery valued at $100.
- 1 horse, 4 milch cows, 4 other cattle, 4 sheep, 5 swine, valued at $421.
- 65 bu. wheat, 100 Indian corn, 85 bu. oats, 10 lbs. wool, 75 bu. Irish potatoes, orchard products valued at $10, 4 gal. wine, 300 lbs. butter, 8 tons hay, 1 bu. grass seed, 5 lbs. flax, 1 bu. flax seed.
- Animals slaughtered valued at $124.
- Estimated value of farm production, $1042.

*Federal census lists Joseph, 48; Susan, 50; John, 18; Joseph, 15; Lydia, 12; Daniel, 11; Jacob, 10; Susan, 9; Samuel, 4; and property valued at $3300.*

### Joseph Zerbe
- 24 acres improved and 2 acres unimproved land valued at $1300.
- Farm machinery valued at $100.
- 1 horse, 1 milch cow, 2 swine, valued at $155.
- 60 bu. wheat, 50 bu. Indian corn, 50 bu. oats, 40 bu. Irish potatoes, 2 bu. sweet potatoes, orchard products valued at $15, 300 lbs. butter, 8 tons hay.
- Animals slaughtered valued at $100.
- Estimated value of farm production, $634.

*Federal census: same as Joseph Zerbe, above, owner of two farms.*

### Henry Brossman
- 90 acres improved and 10 acres unimproved land valued at $6000.
- Farm machinery valued at $255.
- Total wages paid, $125.
- 2 horses, 6 milch cows, 7 other cattle, 5 swine, valued at $835.
- 210 bu. wheat, 200 bu. Indian corn, 200 bu. oats, 25 bu. Irish potatoes, 1 bu. sweet potatoes, orchard products valued at $25, 400 lbs. butter, 20 tons hay, 5 bu. clover seed, 3 bu. grass seed.
- Animals slaughtered valued at $133.
- Estimated value of farm production, $2034.

*Federal census lists Henry, 31; Luzette, 31; John Redcay, 16; Mary Hashuar, a domestic, 18; and property valued at $6000.*

### Benias Schweitzer
- 61 acres improved and 3 acres unimproved land valued at $3200.
- Farm machinery valued at $255.
- Total wages paid, $100.
- 2 horses, 7 milch cows, 3 other cattle, 4 swine, valued at $730.
- 100 bu. wheat, 150 bu. Indian corn, 130 bu. oats, 20 bu. Irish potatoes, orchard products valued at $15, 2 gal. wine, 700 lbs. butter, 20 tons hay, 2 bu. clover seed.
- Animals slaughtered valued at $162.
- Estimated value of farm production, $1741.

*Federal census lists Benias, 40; Susan, 30; Amanda, 8; Ottis, 7; Emary, 5; Nathaniel, 3; Emmaline, 7/12; and property valued at $3200.*

### Solomon Heft
- 32 acres improved and 2 acres unimproved land valued at $1500.
- Farm machinery valued at $100.
- 2 mules, 1 milch cow, 1 other cattle, 2 swine, valued at $363.
- 35 bu. wheat, 40 bu. Indian corn, 80 bu. oats, 25 Irish potatoes, orchard products valued at $25, 200 lbs. butter, 3 tons hay.
- Animals slaughtered valued at $100.
- Estimated value of farm production, $717.

*Federal census lists Solomon, 37; Eliza, 31; Sarah Lutz, 12; and property valued at $1500.*

**Jacob Schlabach**
- 62 acres improved and 5 acres unimproved land valued at $3000.
- Farm machinery valued at $150.
- Total wages paid, $80.
- 2 horses, 6 milch cows, 6 other cattle, 1 sheep, 4 swine, valued at $605.
- 70 bu. wheat, 28 bu. rye, 130 bu. Indian corn, 200 bu, oats, 6 lbs. wool, 50 bu. Irish potatoes, orchard products valued at $10, 600 lbs. butter, 15 tons hay, 1 bu. clover seed.
- Animals slaughtered valued at $226.
- Estimated value of farm production, $1586.

*Federal census lists Jacob, 47; Susan, 56; Catherine, 20; and property valued at $3000.*

**John Kieffer**
- 60 acres improved and 20 acres unimproved land valued at $2500.
- Farm machinery valued at $200.
- Total wages paid, $35.
- 3 horses, 4 milch cows, 5 other cattle, 4 swine, valued at $600.
- 116 bu. wheat, 15 bu. rye, 150 bu. Indian corn, 232 bu. oats, 40 bu. Irish potatoes, orchard products valued at $20, 5 gal. wine, 300 lbs. butter, 20 tons hay, 2 bu. clover seed, 1 bu. grass seed.
- Animals slaughtered valued at $156.
- Estimated value of farm production, $1670.

*Federal census lists John, 61; Catherine, 58; Catherine, 28; Henry, 26; Savilla Coldren, 22, a domestic; and property valued at $2500.*

**Jacob Moseman**
- 15 acres improved and 7 acres unimproved land valued at $1800.
- Farm machinery valued at $50.
- Total wages paid, $8.
- 1 horse, 2 milch cows, 3 swine, valued at $171.
- 30 bu. wheat, 100 bu. Indian corn, 25 bu. Irish potatoes, orchard products valued at $7, 2 gal. wine, 100 lbs. butter, 7 tons hay.
- Animals slaughtered valued at $168.
- Estimated value of farm production, $605.

*Federal census lists Jacob, 75, a minister of the gospel; Philip, 38, a tailor; Barbara Bucher, 36; Mary Bear, 12; Maria Herr; and property valued at $1800. (Jacob and Philip were born in Baden; Barbara was born in Bavaria; Mary was born in Illinois; and Maria was born in Pennsylvania.)*

**William Echternacht**
- 93 acres improved and 10 acres unimproved land valued at $4000.
- Farm machinery valued at $200.
- Total wages paid, $125.
- 5 horses, 6 milch cows, 10 other cattle, 7 swine, valued at $895.
- 200 bu. wheat, 5 bu. rye, 250 bu. Indian corn, 250 bu. oats, 70 bu. Irish potatoes, 3 bu. sweet potatoes, orchard products valued at $8, 3 gal. wine, 500 lbs. butter, 15 tons hay.
- Animals slaughtered valued at $211.
- Estimated value of farm production, $2119.

*Federal census lists William, 44; Levina, 39; Sophia, 16; Albert, 14; Augustine, 12; Mary, 11; Emma, 7; Henry, 3; George, 4/12 (Jan.); and property valued at $4000.*

**John Schlabach**
- 117 acres improved and 8 acres unimproved land valued at $5000.
- Farm machinery valued at $200.
- Total wages paid, $240.
- 7 horses, 8 milch cows, 10 other cattle, 4 swine, valued at $940.
- 150 bu. wheat, 200 bu. Indian corn, 200 bu. oats, 20 bu. Irish potatoes, 2 bu. sweet potatoes, orchard products valued at $10, 500 lbs. butter, 30 tons hay, 2 bu. clover seed, 2 bu. grass seed.
- Animals slaughtered valued at $132.
- Estimated value of farm production, $2166.

*Federal census lists John, 27; Lydia, 33; Maria, 15; Henry Knauer, 15, a farm laborer; Henry Schlebach, 75, retired farmer; and property valued at $5000.*

**Benj. Schlabach**
- 44 acres improved and 12 acres unimproved land valued at $3050.
- Farm machinery valued at $150.
- Total wages paid, $12.
- 2 horses, 5 milch cows, 8 other cattle, 4 sheep, 3 swine, valued at $545.

• 80 bu. wheat, 4 bu. rye, 100 bu. Indian corn, 150 bu. oats, 12 lbs. wool, 30 bu. Irish potatoes, 2 bu. sweet potatoes, orchard products valued at $10, 400 lbs. butter, 15 tons hay, 1 bu. grass seed.
• Animals slaughtered valued at 134.
• Estimated value of farm production, $1313.

*Federal census lists Benj., 35; Magdalene, 48; Lydia, 13; John, 8; Elizabeth Eshelman, 81; and property valued at $3050.*

## Joseph Gehman

• 79 acres improved and 41 acres unimproved land valued at $4000.
• Farm machinery valued at $200.
• Total wages paid, $50.
• 3 horses, 5 milch cows, 5 other cattle, 2 sheep, 4 swine, valued at $601.
• 100 bu. wheat, 10 bu. rye, 200 bu. Indian corn, 200 bu. oats, 10 lbs. wool, 50 bu. Irish potatoes, 2 bu. sweet potatoes, orchard products valued at $25, 300 lbs. butter, 15 tons hay, 2 bu. grass seed.
• Animals slaughtered valued at $148.
• Estimated value of farm production, $1506.

*Federal census lists Joseph 60; Esther, 55; Lydia, 20; Levi Lausch, 12; and property valued at $4000.*

## Elizabeth Gehman

• 60 acres improved land valued at $1000.
• 2 milch cows, 1 other cattle, 3 sheep, 1 swine, valued at $220.
• 18 bu. wheat, 12 bu. Indian corn, 50 bu. oats, 12 bu. Irish potatoes, orchard products valued at $10, 150 lbs. butter, 3 tons hay.
• Animals slaughtered valued at $46.
• Estimated value of farm production, $326.

*Federal census lists Elizabeth, 69; Franklin Knauer, 40, tailor; Mary Knauer, 36; Lewis, 12; George, 6; Catherine, 4; Ida, 1; and property valued at $1000.*

## William Vonneida

• 44 acres improved and 48 acres unimproved land valued at $5000.
• Total wages paid, $70.
• 1 horse, 3 milch cows, 1 other cattle, 2 swine, valued at $231.
• 100 bu. wheat, 60 bu. Indian corn, 130 bu. oats, 25 bu. Irish potatoes, 8 bu. sweet potatoes, 8 gal. wine, 400 lbs. butter, 25 tons hay, forest products valued at $900.
• Animals slaughtered valued at $248.
• Estimated value of farm production, $2138.

*Federal census lists as a possible match, William VonNeida, 49, a miller; Elizabeth, 47; Martha, 24; William, 19, a miller; James, 16; August, 14; Ezra, 10; Robert, 2; and property valued at $17,000.*

## John Hernley

• 58 acres improved and 3 acres unimproved land valued at $3000.
• Farm machinery valued at $150.
• 5 horses, 4 milch cows, 6 other cattle, 3 swine, valued at $560.
• 98 bu. wheat, 2 bu. rye, 100 bu. Indian corn, 250 bu. oats, 15 bu. Irish potatoes, orchard products valued at $15, 35 gal. wine, 400 lbs. butter, 60 tons hay, 7 bu. clover seed, 8 bu. grass seed.
• Animals slaughtered valued at $198.
• Estimated value of farm production, $1407.

*Federal census lists John, 59; Eve, 48; Adam, 20; Rebecca, 18; Julius, 15; Susan, 13; David, 7; and property valued at $3000.*

## Henry Trostle

• 118 acres improved and 3 acres unimproved land valued at $4168.
• Farm machinery valued at $200.
• Total wages paid, $30.
• 2 horses, 5 milch cows, 4 other cattle, 28 sheep, 2 swine, valued at $606.
• 100 bu. wheat, 50 bu. Indian corn, 150 bu. oats, 30 bu. Irish potatoes, orchard products valued at $10, 500 lbs. butter, 28 tons hay.
• Animals slaughtered valued at $217.
• Estimated value of farm production, $1645.

*Federal census lists Henry, 50; Catherine 42; and property valued at $4168.*

## Charles Kramer

• 93 acres improved and 6 acres unimproved land valued at $2000.
• Farm machinery valued at $100.
• Total wages paid, $5.
• 2 horses, 4 milch cows, 8 other cattle, 4 swine, valued at $455.

- 165 bu. wheat, 150 bu. Indian corn, 275 bu. oats, 100 bu. Irish potatoes, 10 bu. sweet potatoes, orchard products valued at $50, 400 lbs. butter, 8 tons hay, 3 bu. clover seed.
- Animals slaughtered valued at $189.
- Estimated value of farm production, $1481.

*Federal census lists Charles, 43, born in Baden; Sophia, 36, born in Würtemberg; Emma, 12; Sophia, 11; Mary, 9; Theresa, 7; Charles, 4; and property valued at $2000.*

### Beneville Pannabecker
- 58 acres improved and 10 acres unimproved land valued at $3500.
- Farm machinery valued at $200.
- Total wages paid, $38.
- 2 horses, 6 milch cows, 5 other cattle, 4 swine, valued at $622.
- 92 bu. wheat, 100 bu. Indian corn, 180 bu. oats, 30 bu. Irish potatoes, 1 bu. sweet potatoes, orchard products valued at $16, 500 lbs. butter, 12 tons hay.
- Animals slaughtered valued at $208.
- Estimated value of farm production, $1466.

*Federal census lists Beneville, 43; Catherine, 43; and property valued at $3500.*

### Michael Weinhold
- 88 acres improved and 18 acres unimproved land valued at $4000.
- Farm machinery valued at $175.
- Total wages paid, $40.
- 3 horses, 6 milch cows, 7 other cattle, 2 sheep, 5 swine, valued at $692.
- 103 bu. wheat, 9 bu. rye, 150 bu. Indian corn, 140 bu. oats, 60 bu. Irish potatoes, orchard products valued at $20, 5 gal. wine, 500 lbs. butter, 20 tons hay, 1 bu. grass seed, forest products valued at $1260.
- Animals slaughtered valued at $160.
- Estimated value of farm production, $2937.

*Federal census lists Michael, 38; Louisa, 32; Mary, 10; Franklin, 9; Amanda, 3; Ida, 1; William, Jr., 39; William, Sr., 72, retired farmer; and property valued at $4000.*

### Peter Hoffman
- 96 acres improved and 18 acres unimproved land valued at $4000.
- Farm machinery valued at $200.
- Total wages paid, $110.
- 6 horses, 7 milch cows, 3 other cattle, 4 sheep, 5 swine, valued at $691.
- 200 bu. wheat, 200 bu. Indian corn, 200 bu. oats, 10 lbs. wool, 50 bu. Irish potatoes, orchard products valued at $10, 500 lbs. butter, 27 tons hay, forest products valued at $1260.
- Animals slaughtered valued at $190.
- Estimated value of farm production, $1991.

*Federal census lists Peter, 74; Christianna, 55; Israel, 21; Isaac, 17; Catherine Weinhold, 25, a domestic; and property valued at $4000.*

### David Hutchinson
- 50 acres improved and 18 acres unimproved land valued at $3500.
- Farm machinery valued at $200.
- Total wages paid, $40.
- 2 horses, 4 milch cows, 6 other cattle, 3 sheep, 5 swine, valued at $469.
- 225 bu. wheat, 150 bu. Indian corn, 150 bu. oats, 25 bu. Irish potatoes, orchard products valued at $150, 400 lbs. butter, 18 tons hay.
- Animals slaughtered valued at $187.
- Estimated value of farm production, $1695.

*Federal census lists David, 38; Margareth, 37; John, 3; Polly Fry, 62; and property valued at $3500.*

### Reuben E. Shober
- 24 acres improved and 16 acres unimproved land valued at $4000.
- Farm machinery valued at $200.
- Total wages paid, $100.
- 2 horses, 5 milch cows, 5 swine, valued at $503.
- 60 bu. wheat, 100 bu. Indian corn, 120 bu. oats, 50 bu. Irish potatoes, 5 bu. sweet potatoes, orchard products valued at $60, 600 lbs. butter, 15 tons hay.
- Animals slaughtered valued at $390.
- Estimated value of farm production, $1582.

*Federal census lists Reuben, 43, a miller; Susan, 23; George, 8; Martin, 6; Ephraim, 3; Margareth, 1; Edward Coldren, 23, laborer; Caroline Sollenberger, 22, a domestic; Susan Shober, 67; and property valued at $9000.*

### Jesse Messner
• 43 acres improved and 3 acres unimproved land valued at $1600.
• Farm machinery valued at $150.
• 2 horses, 3 milch cows, 3 swine, valued at $358.
• 62 bu. wheat, 90 bu. Indian corn, 100 bu. oats, 25 bu. Irish potatoes, orchard products valued at $5, 300 lbs. butter, 5 tons hay, 2 bu. clover seed, 1 bu. grass seed.
• Animals slaughtered valued at $84.
• Estimated value of farm production, $839.
*Federal census lists Jesse, 50; Matilda, 48; Elias, 19; Reese, 14; Matilda, 10; James, 5; and property valued at $1600.*

### Samuel Becker
• 77 acres improved and 13 acres unimproved land valued at $5500.
• Farm machinery valued at $500.
• Total wages paid, $200.
• 4 horses, 3 milch cows, 6 other cattle, 4 swine, valued at $752.
• 240 bu. wheat, 275 bu. Indian corn, 350 bu. oats, 30 bu. Irish potatoes, orchard products valued at $10, 300 lbs. butter, 15 tons hay, 2 bu. clover seed, 2 bu. grass seed.
• Animals slaughtered valued at $156.
• Estimated value of farm production, $1971.
*Federal census lists Samuel, 34; Mary, 35; David, 13; Lydia, 10; Henry, 4; Barbara, 1; and property valued at $5500.*

### Isaac Messner
• 78 acres improved and 4 acres unimproved land valued at $3000.
• Farm machinery valued at $100.
• Total wages paid, $100.
• 2 horses, 5 milch cows, 8 other cattle, 4 swine, valued at $670.
• 85 bu. wheat, 40 bu. rye, 100 bu. Indian corn, 200 bu. oats, 40 bu. Irish potatoes, orchard products valued at $20, 600 lbs. butter, 40 tons hay, 1 bu. grass seed.
• Animals slaughtered valued at $176.
• Estimated value of farm production, $1980.
*Federal census lists Isaac, 61; Harriet, 53; Solomon, 18; Susan, 16; Ellen, 9; and property valued at $3000.*

### George Shup
• 45 acres improved and 57 acres unimproved land valued at $2500.
• Farm machinery valued at $140.
• Total wages paid, $60.
• 2 horses, 4 milch cows, 5 other cattle, 4 swine, valued at $425.
• 50 bu. wheat, 20 bu. rye, 40 bu. Indian corn, 50 bu. oats, 30 bu. Irish potatoes, orchard products valued at $40, 400 lbs. butter, 8 tons hay.
• Animals slaughtered valued at $228.
• Estimated value of farm production, $1737.
*Federal census lists George, 46; Solomon, 28; Mary, 25; George Gensamer, 3; and property valued at $2500.*

### Isaac Good
• 65 acres improved and 10 acres unimproved land valued at $5500.
• Farm machinery valued at $300.
• Total wages paid, $125.
• 4 horses, 5 milch cows, 6 other cattle, 4 swine, valued at $1035.
• 244 bu. wheat, 20 bu. rye, 200 bu. Indian corn, 200 bu. oats, 30 bu. Irish potatoes, orchard products valued at $15, 500 lbs. butter, 25 tons hay, 2 bu. clover seed, forest products valued at $450.
• Animals slaughtered valued at $170.
• Estimated value of farm production, $2157.
*Federal census lists Isaac, 37; Louise, 28; Alvesta, 1; Lydia Steffy, 26, a domestic; and property valued at $5500.*

### Jacob Lessley
• 17 acres improved and 2 acres unimproved land valued at $2800.
• Farm machinery valued at $100.
• Total wages paid, $7.
• 1 horse, 3 milch cows, 1 other cattle, 4 swine, valued at $263.
• 75 bu. wheat, 60 bu. Indian corn, 72 bu. oats, 50 bu. Irish potatoes, orchard products valued $20, 200 lbs. butter, 7 tons hay.
• Animals slaughtered valued at $123.
• Estimated value of farm production, $774
*Federal census lists Jacob, 64, retired; Anna, 54; Catherine, 17; Magdelene, 14; Martin Coldren, 8; and property valued at $2800.*

## Peter Marks

- 125 acres improved and 4 acres unimproved land valued at $8500.
- Farm machinery valued at $350.
- Total wages paid, $50.
- 6 horses, 9 milch cows, 2 other cattle, 7 swine, valued at $1768.
- 400 bu. wheat, 20 bu. rye, 200 bu. Indian corn, 250 bu. oats, 100 bu. Irish potatoes, 2 bu. sweet potatoes, orchard products valued at $10, 200 lbs. butter, 40 tons hay, 1 bu. clover seed, 12 bu. grass seed.
- Animals slaughtered valued at $148.
- Estimated value of farm production, $3551.

*Federal census lists Peter, 41; Emma, 42; Ellen, 16; Jefferson, 15; Franklin, 13; Edward, 12; Catherine, 9; Joseph, 7; Emmaline, 6; John, 3; and property valued at $8500.*

## Jacob Coldren

- 75 acres improved and 3 acres unimproved land valued at $5200.
- Farm machinery valued at $150.
- Total wages paid, $25.
- 2 horses, 4 milch cows, 6 other cattle, 1 sheep, 5 swine, valued at $458.
- 100 bu. wheat, 10 bu. rye, 100 bu. Indian corn, 200 bu. oats, 20 bu. Irish potatoes, orchard products valued at $30; 300 lbs. butter, 16 tons hay, 2 bu. clover seed, 3 bu. grass seed.
- Animals slaughtered valued at $132.
- Estimated value of farm production, $1306.

*Federal census lists Jacob, 49; Lydia, 57; Leman, 12; and property valued at $5200.*

## Isaac Good

- 50 acres improved and 10 acres unimproved land valued at $6600.
- Farm machinery valued at $250.
- Total wages paid, $125.
- 3 horses, 7 milch cows, 3 other cattle, 3 swine, valued at $503.
- 90 bu. wheat, 10 bu. rye, 80 bu. Indian corn, 130 bu. oats, 30 bu. Irish potatoes, 2 bu. sweet potatoes, orchard products valued at $20, 600 lbs. butter, 20 tons hay, 2 bu. clover seed, 1 bu. grass seed.
- Animals slaughtered valued at $210.
- Estimated value of farm production, $1486.

*Federal census lists Isaac, 52; Mary, 55; Lydia, 45; Magdelene, 42; Sarah Steffee, 25, a domestic; Henry Griffith, 1; Daniel Hoshaur, 8; and property valued at $6600.*

## Jacob Burkhart

- 30 acres improved and 2 acres unimproved land valued at $2000.
- Farm machinery valued at $80.
- 2 horses, 3 milch cows, 4 other cattle, 4 swine, valued at $396.
- 55 bu. wheat, 50 bu. Indian corn, 60 bu. oats, 15 bu. Irish potatoes, 4 bu. sweet potatoes, 200 lbs. butter, 8 tons hay, 1 bu. clover seed.
- Animals slaughtered valued at $70.
- Estimated value of farm production, $800.

*Federal census lists Jacob, 43; Rebecca, 40; Lewis, 13; Sarah, 12; Henry, 6; Lydia, 1; and property valued at $2000. Also lists Jacob, 27; Eliza, 29; Eli, 2; Henry Stover, 11; and property valued at $2000.*

## Joseph Beeler

- 42 acres improved and 41 acres unimproved land valued at $2500.
- Farm machinery valued at $50.
- Total wages paid, $30.
- 3 horses, 4 milch cows, 2 other cattle, 1 swine, valued at $495.
- 40 bu. wheat, 7 bu. rye, 50 bu. Indian corn, 60 bu. oats, 75 bu. Irish potatoes, 2 bu. sweet potatoes, orchard products valued at $12, 500 lbs. butter, 15 tons hay.
- Animals slaughtered valued at $122.
- Estimated value of farm production, $1149.

*Federal census lists Joseph, 68; Elizabeth, 72; and property valued at $2500.*

## Jonathan Smith

- 175 acres improved and 175 acres unimproved land valued at $3000.
- Farm machinery valued at $30.
- 1 horse, 2 milch cows, 4 other cattle, valued at $190.
- 12 bu. wheat, 100 bu. Indian corn, 8 bu. Irish potatoes, 200 lbs. butter, 5 tons hay.
- Animals slaughtered valued at $113.
- Estimated value of farm production, $533.

*Federal census lists Jonathan, 61; Elizabeth, 60; Elizabeth, 24; Jestie, 19 (female); Jonathan, 14; and property valued at $3000.*

## Jacob Hurst
- 37 acres improved and 3 acres unimproved land valued at $1600.
- Farm machinery valued at $125.
- 2 horses, 3 milch cows, 1 other cattle, 6 swine, valued at $427.
- 80 bu. wheat, 50 bu. Indian corn, 75 bu. oats, 35 bu. Irish potatoes, orchard products valued at $8, 200 lbs. butter, 8 tons hay, 1 bu. clover seed.
- Animals slaughtered valued at $70.
- Estimated value of farm production, $888.

*Federal census lists Jacob, 34; Rebecca, 33; Maria, 10; Burton, 9; Joseph, 7; David, 4; Abraham, 2; and property valued at $1600.*

## Daniel Glass
- 28 acres improved and 4 acres unimproved land valued at $1100.
- Farm machinery valued at $100.
- 1 horse, 3 milch cows, 3 other cattle, 5 swine, valued at $230.
- 50 bu. wheat, 1 bu. rye, 60 bu. Indian corn, 50 bu. oats, 30 bu. Irish potatoes, 5 bu. sweet potatoes, orchard products valued at $5, 1 gal. wine, 150 lbs. butter, 2 tons hay.
- Animals slaughtered valued at $84.
- Estimated value of farm production, $551.

*Federal census lists Daniel, 42; Lydia, 35; Elizabeth, 13; Lydia, 12; Catherine, 10; Emma, 8; Barton, 3; Franklin, 8/12 (Sept.); and property valued at $1100.*

## William Fox
- 85 acres improved and 85 acres unimproved land valued at $4000.
- Farm machinery valued at $300.
- Total wages paid, $70.
- 6 horses, 7 milch cows, 9 other cattle, 6 swine, valued at $911.
- 250 bu. wheat, 6 bu. rye, 175 bu. Indian corn, 250 bu. oats, 20 bu. Irish potatoes, orchard products valued at $5, 2 gal. wine, 600 lbs. butter, 20 tons hay, 1 bu. clover seed, 2 bu. grass seed.
- Animals slaughtered valued at $208.
- Estimated value of farm production, $2222.

*Federal census lists William, 33; Annetta, 33; Milton, 6; Elizabeth, 4; Aaron, 2; Sarah, 4/12 (Feb.); also Samuel Fox, 67; Elizabeth, 69; and property valued at $4000.*

## Isaac Kieffer
- 25 acres improved and 15 unimproved land valued at $1300.
- Farm machinery valued at $50.
- Total wages paid, $5.
- 1 horse, 2 milch cows, 2 other cattle, 3 swine, valued at $250.
- 62 bu. wheat, 75 bu. Indian corn, 70 bu. oats, 30 bu. Irish potatoes, 250 lbs. butter, 4 tons hay.
- Animals slaughtered valued at $70.
- Estimated value of farm production, $638.

*Federal census lists Isaac, 33; Caroline, 28; Barton, 10; Amanda, 8; Elizabeth, 6; Catherine, 4; William, 1; John, 1/12 (May); and property valued at $1300.*

## John Lorah
- 60 acres improved and 15 acres unimproved land valued at $4000.
- Farm machinery valued at $150.
- 2 horses, 3 milch cows, 2 other cattle, 2 swine, valued at $384.
- 58 bu. wheat, 6 bu. rye, 100 bu. Indian corn, 100 bu. oats, 40 bu. Irish potatoes, 7 bu. sweet potatoes, orchard products valued at $25, 300 lbs. butter, 15 tons hay, 1 bu. grass seed.
- Animals slaughtered valued at $81.
- Estimated value of farm production, $1039.

*Federal census lists John, 34; Susan, 39; David, 11; John, 10; Louisa, 6; Biseler David [David Biseler?], 62; and property valued at $4000.*

## John Youndt
- 23 acres improved land valued at $2500.
- Farm machinery valued at $15.
- 1 horse, 3 milch cows, 2 other cattle, 7 swine, valued at $395.
- 43 bu. wheat, 50 bu. Indian corn, 80 bu. oats, 25 bu. Irish potatoes, orchard products valued at $20, 175 lbs. butter, 10 tons hay.
- Animals slaughtered valued at $165.
- Estimated value of farm production, $925.

*Federal census lists John, 55, a butcher; Sarah, 55; Louisa, 23, a cigar maker; Samuel, 21, a cigar maker; Lydia, 16, a cigar maker; Hetty, 14; Lemar, 12; Mary, 20, a cigar maker; and property valued at $2500.*

## David McColm

- 69 acres improved and 10 acres unimproved land valued at $5000.
- Farm machinery valued at $300.
- Total wages paid, $125.
- 2 horses, 5 milch cows, 10 other cattle, 5 swine, valued at $950.
- 150 bu. wheat, 300 bu. Indian corn, 300 bu. oats, 50 bu. Irish potatoes, 8 bu. sweet potatoes, orchard products valued at $15, 350 lbs. butter, 25 tons hay, 3 bu. clover seed.
- Animals slaughtered valued at $240.
- Estimated value of farm production, $2328.

*Federal census lists David, 59; Lydia, 43; Mary, 15; Sarah, 13; Catherine, 8; David, 3; Catherine Lesher, 68; and property valued at $5000.*

## John Fry

- 65 acres improved and 30 acres unimproved land valued at $3000.
- Farm machinery valued at $375.
- Total wages paid, $125.
- 4 horses, 6 milch cows, 7 other cattle, 2 swine, valued at $819.
- 150 bu. wheat, 350 bu. Indian corn, 300 bu. oats, 25 bu. Irish potatoes, 3 bu. sweet potatoes, orchard products valued at $8, 400 lbs. butter, 30 tons hay, 4 bu. clover seed.
- Animals slaughtered valued at $508.
- Estimated value of farm production, $2580.

*Federal census lists John, 34; Eliza, 28; Susan, 6; David, 4; Catherine, 2; John, 1/12 (May); Lewis Smith, 17, a farm laborer; Maria Sollenberger, 21, serving; and property valued at $3000.*

## George Hoffman

- 100 acres improved and 40 acres unimproved land valued at $6000.
- Farm machinery valued at $400.
- Total wages paid, $190.
- 4 horses, 5 milch cows, 10 other cattle, 5 swine, valued at $1250.
- 225 bu. wheat, 300 bu. Indian corn, 250 bu. oats, 50 bu. Irish potatoes, orchard products valued at $25, 400 lbs. butter, 40 tons hay, forest products valued at $100.
- Animals slaughtered valued at $474.
- Estimated value of farm production, $3257.

*Federal census lists George, 35; Caroline, 34; Reuben, 6; Solomon, 5; Catherine, 3; Peter Hoffman, 41, a farm laborer; John Hoffman, 37, a farm laborer; and property valued at $6000.*

## Nickolas Bowers

- 15 acres improved and 5 acres unimproved land valued at $1400.
- Farm machinery valued at $50.
- 1 horse, 2 milch cows, 2 other cattle, 3 swine, valued at $210.
- 39 bu. wheat, 50 bu. Indian corn, 60 bu. oats, 85 bu. Irish potatoes, orchard products valued at $75, 200 lbs. butter, 3 tons hay.
- Animals slaughtered valued at $126.
- Estimated value of farm production, $647.

*Federal census lists Nickolas, born in Hesse Darmstadt, 46, Catherine (born in Prussia), 52; Catherine, 10; Lydia, 9; and property valued at $1400.*

## John Oecsely

- 14 acres improved and 14 acres unimproved land valued at $2200.
- Farm machinery valued at $40.
- 1 horse, 2 milch cows, 1 other cattle, 3 swine, valued at $173.
- 60 bu. wheat, 8 bu. rye, 80 bu. Indian corn, 50 bu. oats, 25 bu. Irish potatoes, 10 bu. sweet potatoes, orchard products valued at $15; 200 lbs. butter, 5 tons hay.
- Animals slaughtered valued at $70.
- Estimated value of farm production, $584.

*Federal census lists John, 45; Elizabeth, 45; Elizabeth, 11; Anna, 10; Martin, 9; Henry, 8; Mary, 7; George, 4; Louisa, 3; and property valued at $2200.*

## William Wolf

- 20 acres improved and 3 acres unimproved land valued at $1500.
- Farm machinery valued at $30.
- 1 horse, 2 milch cows, 2 other cattle, 4 swine, valued at $180.
- 50 bu. wheat, 2 bu. rye, 40 bu. Indian corn, 90 bu. oats, 25 bu. Irish potatoes, 240 lbs. butter, 7 tons hay, 1 bu. clover seed.
- Animals slaughtered valued at $124.

• Estimated value of farm production, $637.
*Federal census lists William, 55, a shoe maker; Catherine, 47; Jonathan, 10; Jacob, 8; Catherine Marshal, 69; and property valued at $1500.*

## George Dietrich
• 39 acres improved and 5 acres unimproved land valued at $2500.
• Farm machinery valued at $50.
• 2 horses, 4 milch cows, 6 other cattle, 4 swine, valued at $365.
• 149 bu. wheat, 100 bu. Indian corn, 100 bu. oats, 30 bu. Irish potatoes, orchard products valued at $15, 300 lbs. butter, 10 tons hay.
• Animals slaughtered valued at $98.
• Estimated value of farm production, $1054.
*Federal census lists George, 34; Savilla, 23; Mary, 4; and property valued at $2500.*

## Jacob Frankhauser
• 23 acres improved land valued at $2000.
• Farm machinery valued at $40.
• 1 horse, 3 milch cows, 3 other cattle, 3 swine, valued at $187.
• 90 bu. wheat, 100 bu. Indian corn, 130 bu. oats, 40 bu. Irish potatoes, orchard products valued at $50, 250 lbs. butter, 7 tons hay.
• Animals slaughtered valued at $95.
• Estimated value of farm production, $790.
*Federal census lists Jacob, 26; Anna, 24; Milton, 4; Martin, 3; Mary, 1/12 (May); Samuel Strunk, 65, retired farmer; Catherine Frankhauser, 68; and property valued at $2000.*

## Ephraim Lesher
• 41 acres improved land valued at $4000.
• Farm machinery valued at $100.
• 1 horse, 3 milch cows, 3 swine, valued at $237.
• 60 bu. wheat, 25 bu. Indian corn, 100 bu. oats, 20 bu. Irish potatoes, orchard products valued at $50, 400 lbs. butter, 8 tons hay, 3 bu. clover seed, 2 bu. grass seed.
• Animals slaughtered valued at $135.
• Estimated value of farm production, $838.
*Federal census lists Ephraim, 30; Marinda, 25; Michael, 9; and property valued at $4000.*

## Joseph Wenger
• 30 acres improved land valued at $3000.
• Farm machinery valued at $100.
• 2 horses, 4 milch cows, 1 other cattle, 10 swine, valued at $613.
• 60 bu. wheat, 20 bu. Indian corn, 45 bu. oats, 50 bu. Irish potatoes, 5 bu. sweet potatoes, 300 lbs. butter, 8 tons hay.
• Animals slaughtered valued at $224.
• Estimated value of farm production, $1240.
*Federal census lists John, a miller, 44; Susan, 43; Anna, 17; Abraham, 14; Susan, 11; John, 6; Joseph, 2; and property valued at $3000.*

## William Brossman
• 87 acres improved and 15 acres unimproved land valued at $5100.
• Farm machinery valued at $150.
• Total wages paid, $125.
• 4 horses, 6 milch cows, 8 other cattle, 8 swine, valued at $950.
• 300 bu. wheat, 300 bu. Indian corn, 200 bu. oats, 30 bu. Irish potatoes, 8 bu. sweet potatoes, orchard products valued at $20, 20 gal. wine, market gardens produce valued at $30, 400 lbs. butter, 20 tons hay, 1 bu. clover seed, 5 bu. grass seed, 10 lbs. flax, 1 bu. flax seed.
• Animals slaughtered valued at $148.
• Estimated value of farm production, $2405.
*Federal census lists William, 38; Sallie, 35; John, 12; Hannah, 8; Susan, 7; William, 2; and property valued at $5100.*

## Henry Rupp
• 17 acres improved land valued at $2000.
• Farm machinery valued at $40.
• 2 horses, 2 milch cows, 2 swine, valued at $330.
• 30 bu. wheat, 75 bu. Indian corn, 70 bu. oats, 10 bu. Irish potatoes, 100 lbs. butter, 8 tons hay.
• Animals slaughtered valued at $110.
• Estimated value of farm production, $728.
*Federal census lists Henry, 55; Sarah, 49; Henry Wanner, 4; Lydia Laush, 29, a servant; Jacob Bucher; and property valued at $2000.*

## Henry Fry

• 67 acres improved and 6 acres unimproved land valued at $3000.
• Farm machinery valued at $50.
• 1 horse, 5 milch cows, 5 other cattle, 2 swine, valued at $550.
• 100 bu. wheat, 100 bu. Indian corn, 100 bu oats, 25 bu. Irish potatoes, 2 bu. sweet potatoes, 2 gal. wine, 500 lbs. butter, 10 tons hay, 2 lbs. bees wax; 15 lbs. honey.
• Animals slaughtered valued at $208.
• Estimated value of farm production, $1327.

*Federal census lists Henry, 55; Sarah, 49; Henry Wanner, 4; Lydia Laush, 29, a servant; Jacob Bucher, 14; and property valued at $3000.*

## Elias Witmer

• 138 acres improved and 10 acres unimproved land valued at $11,100.
• Farm machinery valued at $450.
• Total wages paid, $350.
• 5 horses, 8 milch cows, 15 other cattle, 8 swine, valued at $1330.
• 645 bu. wheat, 16 bu. rye, 300 bu. Indian corn, 385 bu. oats, 75 bu. Irish potatoes, 5 bu. sweet potatoes, orchard products valued at $15, 700 lbs. butter, 25 tons hay, 5 bu. clover seed, 6 bu. grass seed.
• Animals slaughtered valued at $218.
• Estimated value of farm production, $3549.

*Federal census lists Elias Witwer, 31; Catherine Witwer, 30, Samuel Witwer, 8; Martin Witwer, 5; Joseph Witwer, 2; Anna Witwer, 9/12 (Aug.); Mary Frederic, 26, a servant; John Messner, 18, a farm laborer; and property valued at $11,100.*

## John Oberholtzer

• 46 acres improved and 5 acres unimproved land valued at $6000.
• Farm machinery valued at $200.
• Total wages paid, $125.
• 3 horses, 4 milch cows, 3 other cattle, 4 sheep, 5 swine, valued at $784.
• 200 bu. wheat, 150 bu. Indian corn, 200 bu. oats, 50 bu. Irish potatoes, 2 bu. sweet potatoes, orchard products valued at $20, 400 lbs. butter, 26 tons hay.
• Animals slaughtered valued at $190.
• Estimated value of farm production, $2007.

*Federal census lists John, 50; Lydia, 47; Catherine, 12; Abraham, 9; Barbara, 7; Samuel Davis, 34, a laborer; and property valued at $6000.*

## David Sollenberger

• 71 acres improved and 14 acres unimproved land valued at $4850.
• Farm machinery valued at $100.
• Total wages paid, $100.
• 2 horses, 3 milch cows, 7 other cattle, 12 swine, valued at $515.
• 150 bu. wheat, 200 bu. Indian corn, 200 bu. oats, 50 bu. Irish potatoes, 1 bu. sweet potatoes, orchard products valued at $15, 350 lbs. butter, 15 tons hay.
• Animals slaughtered valued at $148.
• Estimated value of farm production, $1468.

*Federal census lists David, 31; Henriette, 27; Abraham, 9; Ida, 8/12 (Sept.); Susan Lessley, 13; and property valued at $4850.*

## Lazerus Wolf

• 45 acres improved land valued at $2300.
• Farm machinery valued at $50.
• Total wages paid, $10.
• 2 horses, 2 milch cows, 1 other cattle, 4 swine, valued at $315.
• 131 bu. wheat, 100 bu. Indian corn, 260 bu. oats, 25 bu. Irish potatoes, orchard products valued at $5, 200 lbs. butter, 6 tons hay.
• Animals slaughtered valued at $98.
• Estimated value of farm production, $962.

*Federal census lists Lazerus, 28, a butcher; Elizabeth, 23; Ida, 6; Harry, 5; Franklin, 3; Ervin, 1; Michael Wolf, 24, a laborer; Isaac Ulrich, 45, a farm laborer; and property valued at $2300.*

## John Weinhold

• 19 acres improved land valued at $2000.
• Farm machinery valued at $50.
• Total wages paid, $20.
• 1 horse, 3 milch cows, 1 other cattle, 3 swine, valued at $294.

- 56 bu. wheat, 9 bu. rye, 50 bu. Indian corn, 50 bu. oats, 25 bu. Irish potatoes, 3 bu. sweet potatoes, orchard products valued at $30, 3 gal. wine, 300 lbs. butter, 15 tons hay.
- Animals slaughtered valued at $133.
- Estimated value of farm production, $930.

*Federal census lists John, 31; Caroline, 30; Levi, 6; Bennias, 5; Kitty, 3; Sarah, 1; Louisa, 2/12 (Apr.); and property valued at $2000.*

## Isaac K. Beam
- 66 acres improved and 4 acres unimproved land valued at $4500.
- Farm machinery valued at $162.
- Total wages, $75.
- 3 horses, 4 milch cows, 4 other cattle, 4 swine, valued at $679.
- 200 bu. wheat, 50 bu. rye, 250 bu. Indian corn, 400 bu. oats, 100 bu. Irish potatoes, orchard products valued at $125, 500 lbs. butter, 25 tons hay, 5 bu. clover seed, 2 bu. grass seed.
- Animals slaughtered valued at $205.
- Estimated value of farm production, $2341.

*Federal census lists Isaac, 34; Eliza, 27; Charles, 11; Barton, 8; Andrew, 5; Theodore Musselman, 14, a farm laborer; Andrew Gottshall, 13, a farm laborer; and property valued at $4500.*

## Richard Fichthorn
- 37 acres improved land valued at $3400.
- Total wages paid, $50.
- 1 horse, 3 milch cows, 4 swine, valued at $180.
- 60 bu. wheat, 50 bu. Indian corn, 50 bu. oats, 20 bu. Irish potatoes, 9 bu. sweet potatoes, orchard products valued at $10, 200 lbs. butter, 5 tons hay.
- Animals slaughtered valued at $63.
- Estimated value of farm production, $555.

*Federal census lists Richard, 47, a pedlar; Mary, 40; Anna, 13; Cyrus, 11; Catherine, 7; William, 4; and property valued at $3400.*

## John Lessley
- 35 acres improved and 5 acres unimproved land valued at $4000.
- Farm machinery valued at $70.
- Total wages paid, $125.
- 2 horses, 4 milch cows, 3 other cattle, 4 swine, valued at $563.
- 110 bu. wheat, 8 bu. rye, 90 bu. Indian corn, 100 bu. oats, 18 bu. Irish potatoes, 2 bu. sweet potatoes, orchard products valued at $5, 250 lbs. butter, 12 tons hay.
- Animals slaughtered valued at $123.
- Estimated value of farm production, $1225.

*Federal census lists John, 45; Elizabeth, 45; Elizabeth, 11; Anna, 10; Martin, 9; Henry, 8; Mary, 7; George, 4; Louisa, 3; William, 28, a farm laborer; Harriet Coldren, 36, a servant; and property valued at $4000.*

## Jacob Lessley
- 68 acres improved and 12 acres unimproved land valued at $6000.
- Farm machinery valued at $300.
- Total wages paid, $140.
- 4 horses, 5 milch cows, 8 other cattle, 7 swine, valued at $870.
- 270 bu. wheat, 5 bu. rye, 350 bu. Indian corn, 396 bu. oats, 40 bu. Irish potatoes, 4 bu. sweet potatoes, orchard products valued at $25, 3 gal. wine, 600 lbs. butter, 30 tons hay, 3 bu. clover seed, 2 bu. grass seed.
- Animals slaughtered valued at $185.
- Estimated value of farm production, $2607.

*Federal census lists Jacob, 31; Mary, 29; Louisa, 7; Martin, 5; Catherine, 3; Ellen, 1; Jeremiah, 12, a farm laborer; Peggy Messner, 12, and property valued at $6000.*

## John C. Eshelman
- 23 acres improved and 5 acres unimproved land valued at $4000.
- Farm machinery valued at $25.
- Total wages paid, $50.
- 1 horse, 2 milch cows, 2 other cattle, 4 swine, valued at $115.
- 46 bu. wheat, 75 bu. Indian corn, 125 bu. oats, 60 bu. Irish potatoes, orchard products valued at $25, 100 lbs. butter, 6 tons hay, 1 bu. clover seed.
- Animals slaughtered valued at $36.
- Estimated value of farm production, $517.

*Federal census lists John, 28, an innkeeper; Sarah, 19; Laura A., 4; Isaac Stirk, 15, a farm laborer; and property valued at $4000.*

## Amos Schnader

- 32 acres improved and 4 acres unimproved land valued at $2000.
- Farm machinery valued at $30.
- Total wages paid, $20.
- 2 horses, 3 milch cows, 2 other cattle, 6 sheep, 2 swine, valued at $238.
- 30 bu. wheat, 2 bu. rye, 50 bu. Indian corn, 40 bu. oats, 15 bu. Irish potatoes, 1 bu. sweet potatoes, orchard products valued at $10, 300 lbs. butter, 15 tons hay, 2 bu. grass seed.
- Animals slaughtered valued at $109.
- Estimated value of farm production, $919.

*Federal census lists Amos, 63; Catherine, 62; Catherine, 20; Mary, 9; and property valued at $2000.*

## Christian Bleam

- 96 acres improved and 4 acres unimproved land valued at $4000.
- Farm machinery valued at $100.
- Total wages paid, $150.
- 2 horses, 5 milch cows, 5 other cattle, 3 swine, valued at $500.
- 90 bu. wheat, 25 bu. rye, 200 bu. Indian corn, 250 bu. oats, 25 bu. Irish potatoes, 250 lbs. butter, 20 tons hay, 1 bu. clover seed, 1 bu. grass seed.
- Animals slaughtered valued at $160.
- Estimated value of farm production, $1508.

*Federal census lists Christian, 72; Elizabeth, 66; Sarah Coleman, 15, a servant; and property valued at $4000.*

## Baltzer Schnader

- 49 acres improved and 36 acres unimproved land valued at $4500.
- Farm machinery valued at $75.
- Total wages paid, $20.
- 2 horses, 4 milch cows, 4 other cattle, 4 swine, valued at $345.
- 90 bu. wheat, 3 bu. rye, 175 bu. Indian corn, 150 bu. oats, 25 bu. Irish potatoes, 1 bu. sweet potatoes, orchard products valued at $10, 400 lbs. butter, 15 tons hay, 8 bu. clover seed, 2 bu. grass seed.
- Animals slaughtered valued at $151.
- Estimated value of farm production, $1416.

*Federal census lists Baltzer, 38; Elizabeth, 37; Sarah, 14; Davis, 12; Charles, 9; Mary, 5; Martin, 5/12 (Dec.); and property valued at $4500.*

## John H. Good

- 64 acres improved and 20 acres unimproved land valued at $6300.
- Farm machinery valued at $350.
- Total wages paid, $60.
- 3 horses, 5 milch cows, 7 other cattle, 3 sheep, 7 swine, valued at $727.
- 200 bu. wheat, 5 bu. rye, 250 bu. Indian corn, 280 bu. oats, 50 bu. Irish potatoes, 6 bu. sweet potatoes, orchard products valued at $35, 5 gal. wine, 350 lbs. butter, 40 tons hay, 9 bu. clover seed, 5 bu. grass seed, forest products valued at $30.
- Animals slaughtered valued at $227.
- Estimated value of farm production, $2469.

*Federal census lists John, 56; Lavina, 45; Hetty Ann, 22; Martin, 19, a farm laborer; Elizabeth, 17; Israel, 8. Also lists in same household: August Horning, 22; a carpenter; Catherine Horning, 25; Charles Horning, 1; Eliza Good, 48; Sarah Good, 39; and property valued at $6300.*

## Henry Stauffer

- 85 acres improved and 12 acres unimproved land valued at $7000.
- Farm machinery valued at $500.
- Total wages paid, $75.
- 4 horses, 6 milch cows, 8 other cattle, 4 swine, valued at $1136.
- 200 bu. wheat, 300 bu. Indian corn, 200 bu. oats, 50 bu. Irish potatoes, 10 bu. sweet potatoes, orchard products valued at $100, 400 lbs. butter, 30 tons hay.
- Animals slaughtered valued at $207.
- Estimated value of farm production, $2650.

*Federal census lists Henry, 57; Anna, 41; Theodore, 19, a schoolteacher; Mary, 18; Daniel, 16, a farm laborer; Anna, 14; William, 12; Frederic, 10; Margaret, 8; Clara, 5; Benjamin, 4; Alice, 2; Sarah, 5/12 (Dec.); and property valued at $7000.*

## Samuel Shirk

- 40 acres improved land valued at $4000.
- Farm machinery valued at $400.
- Total wages paid, $38.
- 2 horses, 5 milch cows, 6 other cattle, 6 swine, valued at $705.

- 180 bu. wheat, 5 bu. rye, 250 bu. Indian corn, 225 bu. oats, 100 lbs. tobacco, 40 bu. Irish potatoes, 400 lbs. butter, orchard products valued at $50, 20 tons hay.
- Animals slaughtered valued at $161.
- Estimated value of farm production, $2008.

*Federal census lists Samuel, 35; Elizabeth, 34; Barton, 9; William, 6; Henry, 2; and property valued at $4000.*

### Amos Goshen
- 18 acres improved land valued at $1600.
- Farm machinery valued at $60.
- Total wages paid, $15.
- 1 horse, 2 milch cows, 2 swine, valued at $290.
- 55 bu. wheat, 90 bu. Indian corn, 93 bu. oats, 18 bu. Irish potatoes, 150 lbs. butter, 6 tons hay.
- Animals slaughtered valued at $159.
- Estimated value of farm production, $782.

*Federal census lists Amos, 39; Rosanna, 20; Amos, 13; John, 7; George, 5; and property valued at $1600.*

### Jacob Schnader
- 62 acres improved and 15 acres unimproved land valued at $5000.
- Farm machinery valued at $400.
- Total wages paid, $175.
- 3 horses, 6 milch cows, 8 other cattle, 6 swine, valued at $921.
- 200 bu. wheat, 12 bu. rye, 200 bu. Indian corn, 300 bu. oats, 110 bu. Irish potatoes, orchard products valued at $75, market gardens' produce valued at $50, 800 lbs. butter, 30 tons hay.
- Animals slaughtered valued at $198.
- Estimated value of farm production, $2541.

*Federal census lists Jacob, 39; Maria, 38; Obed, 13; Emma, 8; Ervin, 5; Polly More, 38, a servant; and property valued at $5000.*

### Frederick Eshelman
- 82 acres improved and 10 acres unimproved land valued at $6000.
- Farm machinery valued at $200.
- Total wages paid, $25.
- 3 horses, 5 milch cows, 5 other cattle, 6 sheep, 7 swine, valued at $640.
- 200 bu. wheat, 20 bu. rye, 250 bu. Indian corn, 275 bu. oats, 15 lbs. wool, 50 bu. Irish potatoes, orchard products valued at $20, 350 lbs. butter, 17 tons hay, 3 bu. clover seed, 3 bu. grass seed, 30 lbs. flax, 1 bu. flax seed.
- Animals slaughtered valued at $168.
- Estimated value of farm production, $1882.

*Federal census lists Frederick, 68; Elizabeth, 61; Joseph, 23, a farm laborer; Moses, 17, a farm laborer, Elizabeth, 15; and property valued at $6000.*

### Martin Messner
- 79 acres improved land valued at $3000.
- Farm machinery valued at $100.
- Total wages paid, $35.
- 2 horses, 7 milch cows, 5 other cattle, 5 swine, valued at $805.
- 91 bu. wheat, 18 bu. rye, 150 bu. Indian corn, 120 bu. oats, 30 bu. Irish potatoes, orchard products valued at $10, 600 lbs. butter, 20 tons hay.
- Animals slaughtered valued at $177.
- Estimated value of farm production, $1784.

*Federal census lists Martin, 38; Caroline, 27; Lydia, 7; Mary, 5; Eliza, 1/12 (May); Daniel Messner, 74, retired farmer; Catherine, 69; and property valued at $3000.*

### Richard Shiffer
- 75 acres improved and 5 acres unimproved land valued at $5000.
- Farm machinery valued at $200.
- Total wages paid, $12.
- 4 horses, 6 milch cows, 6 other cattle, 4 sheep, 8 swine, valued at $1100.
- 242 bu. wheat, 15 bu. rye, 200 bu. Indian corn, 200 bu. oats, 12 lbs. wool, 50 bu. Irish potatoes, 1 bu. sweet potatoes, orchard products valued at $10, 5 gal. wine, 600 lbs. butter, 30 tons hay.
- Animals slaughtered valued at $465.
- Estimated value of farm production, $2822.

*Federal census lists Richard, 53; Maria, 48; Mary, 19; Israel, 16, a farm laborer; Isaac, 12, a farm laborer; Richard, 10; Susan, 8; John, 7; Abraham, 3; Jacob, 1; and property valued at $5000.*

### Jacob F. Keil
- 29 acres improved land valued at $2600.
- Farm machinery valued at $50.
- 3 horses, 3 milch cows, 2 other cattle, 6 swine, valued at $680.
- 60 bu. wheat, 150 bu. Indian corn, 150 bu. oats, 20 bu. Irish potatoes, 7 bu. sweet potatoes, orchard products valued at $10, 5 gal. wine, 200 lbs. butter, 10 tons hay.
- Animals slaughtered valued at $133.
- Estimated value of farm production, $1330.

*Federal census lists Jacob, 58, a huckster; Eliza, 46; Catherine, 20; James, 12, a farm laborer; Anna, 15; Jacob, 5; George Rishel, 30, a farm laborer; and property valued at $2600.*

### Daniel Weinhold
- 66 acres improved and 12 acres unimproved land valued at $5000.
- Farm machinery valued at $300.
- Total wages paid, $35.
- 4 horses, 3 milch cows, 6 other cattle, 4 swine, valued at $873.
- 200 bu. wheat, 200 bu. Indian corn, 200 bu. oats, 20 bu. Irish potatoes, 2 bu. sweet potatoes, orchard products valued at $20, 4 gal. wine, 400 lbs. butter, 26 tons hay, 7 bu. clover seed, 9 bu. grass seed.
- Animals slaughtered valued at $165.
- Estimated value of farm production, $2208.

*Federal census lists Daniel, 28; Mary, 21; Jacob, 23, a carpenter; Hannah, 50; Samuel, 23, a carpenter, Mary, 21; and property valued at $5000.*

### Joseph Good
- 75 acres improved and 25 acres unimproved land valued at $5000.
- Farm machinery valued at $200.
- Total wages paid, $175.
- 3 horses, 7 milch cows, 6 other cattle, 5 swine, valued at $795.
- 160 bu. wheat, 40 bu. rye, 100 bu. Indian corn, 350 bu. oats, 80 bu. Irish potatoes, orchard products valued at $20, 7 gal. wine, 700 lbs. butter, 25 tons hay, 1 bu. grass seed.
- Animals slaughtered valued at $549.
- Estimated value of farm production, $2499.

*Federal census lists Joseph, 35; Anna, 30; Lewis, 9; Monroe, 6; Esra, 4; Hetty, 1; Henry Harting, 16, a farm laborer; and property valued at $5000.*

### Peter Becker
- 60 acres improved and 5 acres unimproved land valued at $3000.
- Farm machinery valued at $200.
- Total wages paid, $125.
- 5 horses, 4 milch cows, 8 other cattle, 4 swine, valued at $755.
- 160 bu. wheat, 40 bu. rye, 225 bu. Indian corn, 300 bu. oats, 60 bu. Irish potatoes, 3 bu. sweet potatoes, orchard products valued at $40, 500 lbs. butter, 22 tons hay; 1 bu. clover seed, 2 bu. grass seed.
- Animals slaughtered valued at $189.
- Estimated value of farm production, $2046.

*Federal census lists Peter, 35; Hannah, 29; John, 8; Amillia, 6; Elizabeth, 1; John Frankhouser, 15, a farm laborer; Louisa Schlabach, 14, a servant; John Becker, 71, a retired farmer; and property valued at $3000.*

### Jonathan Good
- 64 acres improved and 6 acres unimproved land valued at $7000.
- Farm machinery valued at $573.
- Total wages paid, $125.
- 4 horses, 5 milch cows, 12 other cattle, 11 swine, valued at $750.
- 200 bu. wheat, 7 bu. rye, 200 bu. Indian corn, 200 bu. oats, 40 bu. Irish potatoes, 10 bu. sweet potatoes, orchard products valued at $40, 600 lbs. butter, 35 tons hay, 5 bu. clover seed.
- Animals slaughtered valued at $218.
- Estimated value of farm production, $2327.

*Federal census lists Jonathan, 35; Hetty, 35; Amanda, 15; Mary Ann, 13; Moses, 11; Sarah, 6; and property valued at $7000.*

### David Bender
- 55 acres improved and 7 acres unimproved land valued at $6000.
- Farm machinery valued at $286.
- Total wages paid, $25.
- 3 horses, 5 milch cows, 2 other cattle, 8 swine, valued at $903.

- 125 bu. wheat, 10 bu. rye, 200 bu. Indian corn, 265 bu. oats, 50 bu. Irish potatoes, 3 bu. sweet potatoes, orchard products valued at $30, 500 lbs. butter, 20 tons hay, forest products valued at $25.
- Animals slaughtered valued at $210.
- Estimated value of farm production, $2100.

*Federal census lists David, 39; Eliza, 39; Isaac, 15; Maria, 12; Catherine, 10; David, 7; Eliza, 2; Judith Good, 66; and property valued at $6000.*

### William Reninger
- 55 acres improved and 8 acres unimproved land valued at $6000.
- Farm machinery valued at $639.
- Total wages paid, $70.
- 3 horses, 5 milch cows, 7 other cattle, 4 swine, valued at $705.
- 154 bu. wheat, 42 bu. rye, 300 bu. Indian corn, 300 bu. oats, 8 bu. barley, 75 bu. Irish potatoes, 2 bu. sweet potatoes, orchard products valued at $50; 21 gal. wine, 400 lbs. butter, 24 tons hay, 1 bu. grass seed, 4 lbs. beeswax, 10 lbs. honey.
- Animals slaughtered valued at $193.
- Estimated value of farm production, $2158.

*Federal census lists William, 37; Catherina, 33; Elizabeth, 10; Hetty A., 3; Samuel, 2; Martin Shiffer, 15, a farm laborer; Peter Becker, 69, a retired farmer; Ester Becker, 69; and property valued at $6000.*

### Rudy Brossman
- 25 acres improved and 4 acres unimproved land valued at $2500.
- Farm machinery valued at $100.
- 3 horses, 4 milch cows, 2 other cattle, 4 swine, valued at $535.
- 95 bu. wheat, 100 bu. Indian corn, 100 bu. oats, 50 bu. Irish potatoes, 10 bu. sweet potatoes, orchard products valued at $30, 6 gal. wine, 300 lbs. butter, 12 tons hay, 2 bu. clover seed; 2 bu. grass seed.
- Animals slaughtered valued at $92.
- Estimated value of farm production, $1245.

*Federal census lists Rudy, 30; Elizabeth, 27; John Brossman, 66, a farmer; Hannah, 65; and property valued at $2500.*

### Isaac Reninger
- 89 acres improved and 4 acres unimproved land valued at $7000.
- Farm machinery valued at $533.
- Total wages paid, $140.
- 5 horses, 5 milch cows, 12 other cattle, 11 swine, valued at $1209.
- 300 bu. wheat, 9 bu. rye, 250 bu. Indian corn, 300 bu. oats, 40 bu. Irish potatoes, 5 bu. sweet potatoes, orchard products valued at $50, 5 gal. wine, 600 lbs. butter, 40 tons hay, 5 bu. grass seed.
- Animals slaughtered valued at $180.
- Estimated value of farm production, $3023.

*Federal census lists Isaac, 54; Rebecca, 45; Baltzer, 10; Angeline, 8; Margaret, 6; John, 4; Davis, 2; Mary Ann Fry, 16; Jacob Seifrit, 19, a farm laborer; and property valued at $7000.*

### Jacob Reninger
- 32 acres improved and 7 acres unimproved land valued at $3500.
- Farm machinery valued at $100.
- Total wages paid, $20.
- 3 horses, 3 milch cows, 3 other cattle, 5 swine, valued at $425.
- 109 bu. wheat, 20 bu. rye, 75 bu. Indian corn, 175 bu. oats, 30 bu. Irish potatoes, 2 bu. sweet potatoes, orchard products valued at $20, 200 lbs. butter, 10 tons hay, 2 bu. clover seed.
- Animals slaughtered valued at $109.
- Estimated value of farm production, $1104.

*Federal census lists Jacob, 59, a blacksmith; Catherine, 55; Edward 22, a farm laborer; Susan, 14; Lamanda, 10; and property valued at $3500.*

### Cyrus Messner
- 44 acres improved and 6 acres unimproved land valued at $4000.
- Farm machinery valued at $200.
- 2 horses, 4 milch cows, 4 other cattle, 5 swine, valued at $375.
- 70 bu. wheat, 100 bu. Indian corn, 100 bu. oats, 20 bu. Irish potatoes, 2 bu. sweet potatoes, orchard products valued at $20, 400 lbs. butter, 10 tons hay, 3 bu. clover seed, 2 bu. grass seed.
- Animals slaughtered valued at $105.
- Estimated value of farm production, $870.

*Federal census lists Cyrus, 56, a carpenter; Lydia, 52; Lydia A., 15; Heber, 13; Levi, 10; Tobias Missner, 25, a laborer; Lydia, 25; and property valued at $4000.*

## Henry Steiner
- 37 acres improved land valued at $2400.
- Farm machinery valued at $75.
- 1 horse, 2 milch cows, 2 other cattle, 3 swine, valued at $228.
- 116 bu. wheat, 40 bu. Indian corn, 140 bu. oats, 50 lbs. tobacco, 40 bu. Irish potatoes, 2 bu. sweet potatoes, 100 lbs. butter, 8 tons hay, 3 bu. clover seed.
- Animals slaughtered valued at $45.
- Estimated value of farm production, $793.

*Federal census lists Henry, 45; Louisa, 39; Elizabeth, 8; Monroe Schlabach, 7; and property valued at $2400.*

## Rudolph Killian
- 90 acres improved and 3 acres unimproved land valued at $6000.
- Farm machinery valued at $200.
- Total wages paid, $90.
- 3 horses, 5 milch cows, 7 other cattle, 4 swine, valued at $743.
- 220 bu. wheat, 175 bu. Indian corn, 350 bu. oats, 30 bu. Irish potatoes, 5 bu. sweet potatoes, orchard products valued at $5, 250 lbs. butter, 25 tons hay, 4 bu. clover seed, 20 bu. grass seed.
- Animals slaughtered valued at $262.
- Estimated value of farm production, $2217.

*Federal census lists Rudolph, 27; Catherine, 30; Wayne, 9; Diller, 7; Kuntzer, 5; Monroe, 8/12 (Sept.); Henry Pierce, 12, a farm laborer; and property valued at $6000.*

## Samuel Lessley
- 27 acres improved land valued at $3000.
- Farm machinery valued at $100.
- Total wages paid, $25.
- 1 horse, 2 milch cows, 2 other cattle, 2 swine, valued at $383.
- 163 bu. wheat, 125 bu. Indian corn, 150 bu. oats, 5 bu. Irish potatoes, orchard products valued at $25, 300 lbs. butter, 7 tons hay.
- Animals slaughtered valued at $136.
- Estimated value of farm production, $1108.

*Federal census lists Samuel, 61, a blacksmith; Margaret, 67; Rebecca, 27; Henry, 5; and property valued at $3000.*

## Christian Miller
- 25 acres improved and 18 acres unimproved land valued at $2000.
- Farm machinery valued at $40.
- Total wages paid, $90.
- 1 horse, 4 milch cows, 3 other cattle, 4 swine, valued at $433.
- 72 bu. wheat, 90 bu. Indian corn, 133 bu. oats, 40 bu. Irish potatoes, 400 lbs. butter, 9 tons hay.
- Animals slaughtered valued at $150.
- Estimated value of farm production, $1077.

*Federal census lists Christian, 43, a carpenter; Susan, 45; Henry, 16, a farm laborer; Catherine, 68; and property valued at $2000.*

## Adam Hornberger
- 31 acres improved and 1 acre unimproved land valued at $3000.
- Farm machinery valued at $50.
- Total wages paid, $70.
- 1 horse, 3 milch cows, 1 sheep, 5 swine, valued at $387.
- 84 bu. wheat, 150 bu. Indian corn, 70 bu. oats, 5 lbs. wool, 75 bu. Irish potatoes, 2 bu. sweet potatoes, orchard products valued at $15, 3 gal. wine, 300 lbs. butter, 8 tons hay.
- Animals slaughtered valued at $159.
- Estimated value of farm production, $1077.

*Federal census lists Adam, 34, a wheelwright; Catherine, 34; Hetty Ann, 7; Emmaline, 5; Milton, 4; Joseph, 1; James Fox, 13; and property valued at $3000.*

## Isaac Burkhart
- 105 acres improved and 20 acres unimproved land valued at $7560.
- Farm machinery valued at $300.
- Total wages paid, $150.
- 5 horses, 9 milch cows, 10 other cattle, 11 swine, valued at $1130.
- 200 bu. wheat, 10 bu. rye, 200 bu. Indian corn, 500 bu. oats, 75 bu. Irish potatoes, orchard products valued at $20, 800 lbs. butter, 40 tons hay, 6 bu. clover seed, 3 bu. grass seed, forest products valued at $15.
- Animals slaughtered valued at $376.

• Estimated value of farm production, $3142.
*Federal census lists Isaac, 57; Ester, 43; John, 27, a carpenter; Anna, 18; Catherine, 1; Martin, 14; David, 11; Amos, 3; Mary, 2; Rebecca, 1/12 (May); Daniel Good, 62, a weaver; and property valued at $7560.*

### Peter Musser
• 87 acres improved and 7 acres unimproved land valued at $5100.
• Farm machinery valued at $200.
• Total wages paid, $40.
• 4 horses, 6 milch cows, 7 other cattle, 5 swine, valued at $805.
• 200 bu. wheat, 14 bu. rye, 150 bu. Indian corn, 200 bu. oats, 100 bu. Irish potatoes, 5 bu. sweet potatoes, orchard products valued at $15, 500 lbs. butter, 25 tons hay, 4 bu. clover seed, 1 bu. grass seed.
• Animals slaughtered valued at $195.
• Estimated value of farm production, $2116.
*Federal census lists Peter, 50; Sarah, 50; Israel, 19, a farm laborer; Mary, 17; Sarah, 10; Franklin, 21, a laborer; Mary, 20; Harvey, 5/12 (Dec.); and property valued at $5100.*

### Jacob Musser
• 60 acres improved and 7 acres unimproved land valued at $4370.
• Farm machinery valued at $200.
• Total wages paid, $20.
• 3 horses, 5 milch cows, 3 other cattle, 4 swine, valued at $682.
• 150 bu. wheat, 12 bu. rye, 150 bu. Indian corn, 240 bu. oats, 150 bu. Irish potatoes, orchard products valued at $20, 500 lbs. butter, 26 tons hay, 6 bu. clover seed, 2 bu. grass seed.
• Animals slaughtered valued at $217.
• Estimated value of farm production, $1939.
*Federal census lists Jacob, 48; Magdelene, 44; Susan, 22; Solomon, 17, a farm laborer; Lydia, 14; Jacob, 12; and property valued at $4370.*

### Henry Vonnieda
• 36 acres improved and 7 acres unimproved land valued at $3000.
• Farm machinery valued at $200.
• Total wages paid, $325.
• 2 horses, 4 milch cows, 4 other cattle, 7 swine, valued at $822.
• 120 bu. wheat, 6 bu. rye, 50 b. Irish potatoes, 5 bu. sweet potatoes, orchard products valued at $50, 2 gal. wine, 500 lbs. butter, 20 tons hay, 2 bu. clover seed.
• Animals slaughtered valued at $442.
• Estimated value of farm production, $1955.
*Federal census lists Henry, 50, a miller; Sarah, 36; Emmaline, 18; Isabella, 12; Adam, 10; Barton, 6; Nancy, 3; Elizabeth Grill, 50, domestic servant; Isaac Walter, a miller, 23; Joseph Musser, a farm laborer, 28; and property valued at $3000.*

### Moses Horning
• 94 acres improved and 8 acres unimproved land valued at $8000.
• Farm machinery valued at $200.
• Total wages paid, $65.
• 3 horses, 2 mules and asses; 6 milch cows, 8 other cattle, 9 swine, valued at $1000.
• 180 bu. wheat, 75 bu. rye, 200 bu. Indian corn, 200 bu. oats, 250 bu. Irish potatoes, 7 bu. sweet potatoes, orchard products valued at $10, 700 lbs. butter, 30 tons hay, 2 bu. clover seed, 2 bu. grass seed.
• Animals slaughtered valued at $228.
• Estimated value of farm production, $2606.
*Federal census lists Moses, 40; Lavina, 37; Elizabeth, 16; Joseph, 14, a farm laborer; David, 12; Lydia, 10; Franklin, 8; Benjamin, 5; Israel, 2; and property valued at $8000.*

### Maria Frankhauser
• 46 acres improved and 9 acres unimproved land valued at $2500.
• Farm machinery valued at $150.
• Total wages paid, $128.
• 2 mules and asses; 6 milch cows; 7 other cattle, 3 swine, valued at $655.
• 122 bu. wheat, 175 bu. Indian corn, 230 bu. oats, 25 bu. Irish potatoes, 2 bu. sweet potatoes, 520 lbs. butter, 15 tons hay, 1 bu. clover seed.
• Animals slaughtered valued at $140.
• Estimated value of farm production, $1606.
*Federal census lists Maria, 36; Freemont, 13, a farm laborer; Anna, 10; Lot, 7; John, 5; Sherman, 2; and property valued at $2500.*

## Peter B. Good

- 65 acres improved and 61 acres unimproved land valued at $7335.
- Farm machinery valued at $150.
- Total wages paid, $200.
- 3 horses, 4 milch cows, 5 swine, valued at $715.
- 95 bu. wheat, 200 bu. Indian corn, 150 bu. oats, 25 bu. Irish potatoes, orchard products valued at $5, 400 lbs. butter, 15 tons hay, 2 bu. clover seed.
- Animals slaughtered valued at $664.
- Estimated value of farm production, $7112.

*Federal census lists Peter, 46, a miller; Hannah, 43; Mary, 18; Leman, 17; Emmaline, 14; John Werner, 21, a miller; Elias Gehman, 17, farm laborer; Mary Redcay, 41, keeping house; Edward Redcay, 9; and property valued at $11,835.*

## Simon Kind

- 38 acres improved and 21 acres unimproved land valued at $2000.
- Farm machinery valued at $160.
- Total wages paid, $5.
- 2 horses, 6 milch cows, 4 other cattle, 2 sheep, 4 swine, valued at $518.
- 80 bu. wheat, 140 bu. Indian corn, 100 bu. oats, 12 lbs. wool, 25 bu. Irish potatoes, 2 bu. sweet potatoes, orchard products valued at $10, 400 lbs. butter, 15 tons hay, 2 bu. clover seed, 10 lbs. flax.
- Animals slaughtered valued at $180.
- Estimated value of farm production, $1347.

*Federal census lists Simon, 49; Mary, 57; Amanda, 15; Sarah Trostel, 18, a servant; and property valued at $2000.*

# Brecknock Township—1880

*The 1880 agricultural census was very thorough in that each farmer had to answer 100 questions about his farm. Of course, some of the questions did not pertain to his farm. In order to more fully appreciate the contents of the 1880 agricultural census, the contents of those questions are herein given for reference while using the entries for each farmer.*

*It should be noted that here we have had to match the agricultural census with the Federal census without the aid of a value on their property. It was very easy in most cases, as the same enumerator, Henry Stauffer, did both censuses simultaneously and in the cases of where the first names do not match, the families were listed in the same order on both censuses.*

## Henry Stauffer

- Owner
- 60 tilled acres, 8 acres permanent meadows, 6 acres woodland forests, 6 acres unimproved land valued at $8000.
- Farm machinery valued at $600; livestock, $500; fences, $20; fertilizers, $40; wages for farm labor, $200; weeks of hired labor, 30; estimated value of all farm productions, $900.
- Acres of grass mown, 18; not mown, 6; 25 tons of hay harvested.
- 4 horses, 1 milch cow, 5 other cattle, 1 calf dropped, 4 cattle sold living, 1 cattle slaughtered, 300 lbs. butter, 8 swine, 40 poultry in barnyard, 300 dozen eggs produced.
- 10 acres Indian corn, 300 bu. Indian corn, 6 acres oats, 200 bu. oats, 8 acres wheat, 200 bu. wheat, 1/2 acre Irish potatoes, 25 bu. Irish potatoes, 1/8 acre sweet potatoes, 10 bu. sweet potatoes, 5 acres tobacco, 1300 lbs. tobacco.
- 1 acre with 50 apple trees, 30 bu. apples; total value of orchard products, $15.
- 10 cords of wood cut valued at $10.

*Federal census lists Henry, 67; Anna, 50 (wife); Anna C., 24, a teacher; William H., 21; Frederic A., 20, a teacher; Mag. A., 18; Benjamin C., 14; Sarah L., 10.*

## Michael Witwer

- Owner
- 80 tilled acres, 10 acres permanent meadows, 20 acres woodland forests, 19 acres unimproved land valued at $6500.
- Farm machinery valued at $400; livestock, $500; fences, $20; fertilizers, $100; wages for farm labor, $200; weeks of hired labor, 30; estimated value of all farm productions, $900.
- Acres of grass mown, 30; not mown, 10; 25 tons of hay harvested.
- 4 horses, 1 milch cow; 5 other cattle, 7 calves dropped, 2 cattle slaughtered, 200 lbs. butter, 7 swine, 30 poultry in barnyard, 150 dozen eggs produced.
- 17 acres Indian corn, 600 bu. Indian corn, 10 acres oats, 300 bu. oats, 20 acres wheat, 300 bu. wheat, 2 acres Irish potatoes, 150 bu. Irish potatoes.
- 3 acres with 100 apple trees, 25 bu. apples; total value of orchard products, $12.
- 10 cords of wood cut valued at $30.

*Federal census lists Michael, 52; Eliza, 43 (wife); Eliza, 21; Anna, 19; James, 17; Lydia, 16; Hattie, 14; Michael, 13; William, 8; Lizzie, 10; Peter, 6, Mary, 3.*

## Anthony Good

- Owner
- 60 tilled acres, 12 acres permanent meadows, 40 acres woodland forests, 18 acres unimproved land valued at $4000.
- Farm machinery valued at $300; livestock, $400; fences, $10; fertilizers, $100; wages for farm labor, $100; weeks of hired labor, 25; estimated value of all farm productions, $600.
- Acres of grass mown, 20; not mown, 9; 18 tons of hay harvested.
- 3 horses, 5 milch cows, 5 other cattle, 4 calves dropped, 1 cattle slaughtered, 200 lbs. butter, 6 swine, 60 poultry in barnyard, 140 dozen eggs produced.
- 9 acres Indian corn, 200 bu. Indian corn, 1 acre oats, 150 bu. oats, 9 acres wheat, 140 bu. wheat, 3/4 acre Irish potatoes, 40 bu. Irish potatoes, 1/16 acre sweet potatoes, 6 bu. sweet potatoes, 1/4 acre tobacco, 300 lbs. tobacco.
- 2 acres with 50 apple trees, 10 bu. apples; total value of orchard products, $5.
- 10 cords of wood cut valued at $12.

*Federal census lists Anthony, 69; Rachel, 56 (wife); William P., 26; Caroline B., 21; Anna M., 17; Edward H., 16; Jacob Good, 40, boarder (farm laborer); Catherine Yohn, 90, widow (mother-in-law).*

## Samuel Reifsnyder

- Owner
- 50 tilled acres, 5 acres permanent meadows, 12 acres woodland forests, valued at $4500.
- Farm machinery valued at $400; livestock, $400; fences, $10; fertilizers, $60; wages for farm labor, $150; weeks of hired labor, 30; estimated value of all farm productions, $725.
- Acres of grass mown, 16; not mown, 4; 25 tons of hay harvested.
- 3 horses, 5 milch cows, 1 other cattle, 5 calves dropped, 1 cattle slaughtered, 250 lbs. butter, 10 swine, 50 poultry in barnyard, 150 dozen eggs produced.
- 10 acres Indian corn, 250 bu. Indian corn, 7 acres oats, 200 bu. oats, 7 acres wheat, 165 bu. wheat, 1 acre Irish potatoes, 40 bu. Irish potatoes, 1/4 acre tobacco, 375 lbs. tobacco.
- 1 acre with 30 apple trees, 10 bu. apples; total value of orchard products, $6.
- 14 cords of wood cut valued at $14.

*Federal census lists Samuel, 43; Caroline, 39 (wife); Charles B., 16; Sallie C., 14; Isaac M., 12; William H., 10; Irwin, 6.*

## Isaac Becker

- Owner
- 16 tilled acres, 1 acre permanent meadows, valued at $2000.
- Farm machinery valued at $30; livestock, $100; fences, $15; fertilizers, $10; wages for farm labor, $6; weeks of hired labor, 1; estimated value of all farm productions, $200.
- Acres of grass mown, 6; not mown, 1; 4 tons of hay harvested.
- 1 horse, 2 milch cows, 2 calves dropped, 2 cattle sold living, 150 lbs. butter, 2 swine, 30 poultry in barnyard, 250 dozen eggs produced.
- 2 acres Indian corn, 50 bu. Indian corn, 2 acres oats, 160 bu. oats, 2 acres wheat, 40 bu. wheat, 1/2 acre Irish potatoes, 20 bu. Irish potatoes.
- 4 cords of wood cut valued at $5.

*Federal census lists Isaac, 56; Sally, 63 (wife).*

## Christian and Benjamin Good

- Owner
- 30 tilled acres, 5 acres permanent meadows, 4 acres woodland forests, 1 acre unimproved land, valued at $3000.
- Farm machinery valued at $200; livestock, $200; fences, $20; fertilizers, $30; wages for farm labor, $200; weeks of hired labor, 30; estimated value of all farm productions, $450.
- Acres of grass mown, 14; not mown, 3; 15 tons of hay harvested.
- 1 horse, 5 milch cows, 1 other cattle, 5 calves dropped, 5 cattle sold living, 1 cattle slaughtered, 250 lbs. butter, 4 swine, 60 poultry in barnyard, 300 dozen eggs produced.
- 6 acres Indian corn, 250 bu. Indian corn, 6 acres oats, 160 bu. oats, 4 acres wheat, 90 bu. wheat, 1/2 acre Irish potatoes, 50 bu. Irish potatoes.
- 2 acres with 67 apple trees, 20 bu. apples; total value of orchard products, $11.
- 7 cords of wood cut valued at $10.

*Federal census lists Christian, 44; Benjamin, 48 (brother); Elisabeth, 49 (sister).*

## Solomon Ott

- Owner
- 55 tilled acres, 8 acres permanent meadows, 3 acres woodland forests, 1 acre unimproved land, valued at $5000.
- Farm machinery valued at $500; livestock, $400; fences, $20; fertilizers, $50; wages for farm labor, $300; weeks of hired labor, 25; estimated value of all farm productions, $700.
- Acres of grass mown, 20; not mown, 5; 30 tons of hay harvested.
- 2 horses, 5 milch cows, 4 other cattle, 5 calves dropped, 3 cattle sold living, 1 cattle slaughtered, 300 lbs. butter, 6 swine, 40 poultry in barnyard, 300 dozen eggs produced.
- 9 acres Indian corn, 370 bu. Indian corn, 6 acres oats, 200 bu. oats, 1 acre rye, 18 bu. rye, 7 acres wheat, 155 bu. wheat, 3/4 acres Irish potatoes, 40 bu. Irish potatoes, 1/2 acre tobacco, 300 lbs. tobacco.
- 3 acres with 50 apple trees, 30 bu. apples; total value of orchard products, $15.
- 6 cords of wood cut valued at $7.

*Federal census lists Solomon, 49; Lydia, 36 (wife); Mary, 15; Lizzie, 12; Emma, 5; Veronica, 3; Veronica, 71 (widowed mother); Steffe Catherine [Catherine Steffe?], 13 (step-dau.).*

## Richard Boehm

- Owner
- 12 tilled acres, 6 acres permanent meadows, 6 acres woodland forests, valued at $2000.
- Farm machinery valued at $200; livestock, $300; fences, $10; fertilizers, $50; wages for farm labor, $50; weeks of hired labor, 25; estimated value of all farm productions, $400.
- Acres of grass mown, 18; not mown, 24; 26 tons of hay harvested; 3 bu. clover seed.

- 2 horses, 1 milch cow, 5 other cattle, 4 calves dropped, 2 cattle sold living, 1 cattle slaughtered, 200 lbs. butter, 4 swine, 30 poultry in barnyard, 150 dozen eggs produced.
- 6 acres Indian corn, 150 bu. Indian corn, 4 acres oats, 100 bu. oats, 6 acres wheat, 100 bu. wheat, 1/4 acre Irish potatoes, 20 bu. Irish potatoes.
- 2 acres with 50 apple trees.
- 5 cords of wood cut valued at $5.

*Federal census lists Richard, 50 (widower); Susan, 25.*

## Henry G. Good

- Owner
- 34 tilled acres, 3 acres permanent meadows, valued at $3500.
- Farm machinery valued at $400; livestock, $200; fences, $10; fertilizers, $90; wages for farm labor, $30; weeks of hired labor, 10; estimated value of all farm productions, $300.
- Acres of grass mown, 8; 14 tons of hay harvested.
- 2 horses, 3 milch cows, 3 calves dropped, 3 cattle sold living, 1 cattle slaughtered, 200 lbs. butter, 3 swine, 18 poultry in barnyard, 25 dozen eggs produced.
- 4 acres Indian corn, 150 bu. Indian corn, 5 acres oats, 214 bu. oats, 3 acres rye, 42 bu. rye, 6 acres wheat, 103 bu. wheat, 3/4 acre Irish potatoes, 40 bu. Irish potatoes, 1/10 acre sweet potatoes, 8 bu. sweet potatoes.
- 1 acre with 30 apple trees, 5 bu. apples; total value of orchard products, $2.

*Federal census lists Henry, 34; Sarah, 25 (wife); Susanna, 6; Jacob, 4; Samuel, 6/12 (Nov.).*

## David Horning

- Rents for shares of products.
- 37 tilled acres, 2 acres permanent meadows, valued at $3500.
- Farm machinery valued at $200; livestock, $200; fences, $50; fertilizers, $40; wages for farm labor, $25; weeks of hired labor, 10; estimated value of all farm productions, $300.
- Acres of grass mown, 8; not mown, 4; 8 tons of hay harvested; 1 bu. clover seed.
- 1 horse, 3 milch cows, 1 other cattle, 3 calves dropped, 3 cattle sold living, 1 cattle slaughtered, 200 lbs. butter, 3 swine, 16 poultry in barnyard, 50 dozen eggs produced.
- 7 acres Indian corn, 200 bu. Indian corn, 6 acres oats, 200 bu. oats, 5 acres wheat, 96 bu. wheat, 1/2 acre Irish potatoes, 40 bu. Irish potatoes, 1/10 acre sweet potatoes, 8 bu. sweet potatoes.
- 2 acres with 7 apple trees, 10 bu. apples; total value of orchard products, $2.

*Federal census lists David, 32; Fanny, 23 (wife).*

## Rebecca Renninger

- Owner
- 7 tilled acres, 1 acre permanent meadows, valued at $1200.
- Farm machinery valued at $50; livestock, $100; fences, $25; fertilizers, $10; wages for farm labor, $10; weeks of hired labor, 2; estimated value of all farm productions, $150.
- Acres of grass mown, 2; not mown, 1; 4 tons of hay harvested.
- 1 milch cow, 1 other cattle, 2 calves dropped, 2 cattle sold living, 75 lbs. butter, 2 swine, 30 poultry in barnyard, 100 dozen eggs produced.
- 2 acres Indian corn, 50 bu. Indian corn, 4 acres rye, 5 bu. rye, 1 acre wheat, 23 bu. wheat, 1/4 acre Irish potatoes, 20 bu. Irish potatoes.
- 1/4 acre with 25 apple trees, 2 bu. apples, 1/10 acre with 20 peach trees; total value of orchard products, $1.

*Federal census lists Rebecca, 51; Baltzer, 20; Angeline, 18; Margaret, 16; John A., 14; Davis, 11; Tobias, 9.*

## Baltzer Schneder

- Owner
- 50 tilled acres, 2 acres permanent meadows, 12 acres woodland forests, valued at $3000.
- Farm machinery valued at $700; livestock, $700; fences, $100; fertilizers, $60; wages for farm labor, $100; weeks of hired labor, 25; estimated value of all farm productions, $100.
- Acres of grass mown, 15; not mown, 2; 12 tons of hay harvested.
- 3 horses, 7 milch cows, 1 other cattle, 7 calves dropped, 3 cattle sold living, 1 cattle slaughtered, 300 lbs. butter, 7 swine, 40 poultry in barnyard, 200 dozen eggs produced.
- 8 acres Indian corn, 300 bu. Indian corn, 6 acres oats, 200 bu. oats, 6 acres rye, 5 bu. rye, 7 acres wheat, 192 bu. wheat, 1 acre Irish potatoes, 35 bu. Irish potatoes, 2 bu. sweet potatoes.
- 2 acres with 50 apple trees, 10 bu. apples, 1/10 acre with 20 peach trees, 1 bu. peaches; total value of orchard products, $10.
- 135 cords of wood cut valued at $122.

*Federal census lists Baltzer, 48; Lizzie, 47 (wife); Sarah B., 25; Davis B., 23; Charles B., 19; Mary B., 16; Amos B., 4; Emma Good, 6 (granddaugher).*

## Mariah Frankhauser

- Owner
- 60 tilled acres, 10 acres permanent meadows, 1/2 acre woodland forests, valued at $3000.
- Farm machinery valued at $700; livestock, $600; fences, $100; fertilizers, $50; wages for farm labor, $50; weeks of hired labor, 10; estimated value of all farm productions, $500.
- Acres of grass mown, 15; now mown, 3; 15 tons of hay harvested.
- 3 horses, 5 milch cows, 2 other cattle, 5 calves dropped, 1 cattle sold living, 1 cattle slaughtered, 500 lbs. butter, 2 fleeces, 6 lbs. wool, 4 sheep sold living, 4 swine, 50 poultry in barnyard, 500 dozen eggs produced.
- 8 acres Indian corn, 250 bu. Indian corn, 6 acres oats, 200 bu. oats, 1/2 acre rye, 5 bu. rye, 6 acres wheat, 100 bu. wheat, 1 acre Irish potatoes, 65 bu. Irish potatoes, 1/2 acre tobacco, 250 lbs. tobacco.
- 1 acre with 45 apple trees, 10 bu. apples; total value of orchard products, $10.
- Market gardens valued at $76.
- 5 cords of wood cut valued at $5.

*Federal census lists Mary, 46; Lot W., 17, son, farm laborer; Jno. W., 15 (son), farm laborer, Sherman W., 13 (son), farm laborer; Samuel Beaver, 27, farm laborer; Anna E. Beaver, 21, wife; Nora, 3/4 (Apr.).*

## Martin Messner

- Owner
- 50 tilled acres, 8 acres permanent meadows, 4 acres woodland forests, valued at $3000.
- Farm machinery valued at $200; livestock, $400; fences, $100; fertilizers, $50; wages for farm labor, $50; weeks of hired labor, 10; estimated value of all farm productions, $500.
- Acres of grass mown, 15; not mown, 4; 15 tons of hay harvested.
- 2 horses, 5 milch cows, 6 other cattle, 5 calves dropped, 250 lbs. butter, 4 swine, 40 poultry in barnyard, 200 dozen eggs produced.
- 4 acres Indian corn, 250 bu. Indian corn, 6 acres oats, 200 bu. oats, 1/2 acre rye, 5 bu. rye, 6 acres wheat, 60 bu. wheat, 1/4 acre Irish potatoes, 25 bu. Irish potatoes.
- 1/4 acres with 25 apple trees.
- 10 cords of wood cut valued at $15.

*Federal census lists Martin, 40; Caroline, 37 (wife); Lizzie, 15; Mary, 13; Eliza, 10; Amamda, 7; Martin, 5.*

## Frederic Eshleman

- Owner
- 60 tilled acres, 4 acres permanent meadows, 6 acres woodland forests, 10 acres unimproved land, valued at $5000.
- Farm machinery valued at $700; livestock, $400; fences, $50; wages for farm labor, $200; weeks of hired labor, 20; estimated value of all farm productions, $800.
- Acres of grass mown, 15; not mown, 15; 30 tons of hay harvested.
- 4 horses, 4 milch cows, 5 other cattle, 4 calves dropped, 200 lbs. butter, 5 sheep sold living, 10 swine, 50 poultry in barnyard, 200 dozen eggs produced.
- 11 acres Indian corn, 200 bu. Indian corn, 8 acres oats, 300 bu. oats, 1 acre rye, 11 bu. rye, 12 acres wheat, 170 bu. wheat, 1/4 acre Irish potatoes, 40 bu. Irish potatoes, 1 acre tobacco, 1100 lbs. tobacco.
- 1 acre with 60 apple trees; 2 bu. apples; total value of orchard products, $2.
- 20 cords of wood cut valued at $20.

*Federal census lists Frederic, 77; Elizabeth, 71 (wife); Joseph, 34; Moses, 24; Lizzie, 22.*

## Isaac Walter

- Owner
- 60 tilled acres, 12 acres permanent meadows, 5 acres woodland forests, 10 acres unimproved land, valued at $5000.
- Farm machinery valued at $600; livestock, $600; fences, $100; fertilizers, $100; wages for farm labor, $200; weeks of hired labor, 30; estimated value of all farm productions, $600.
- Acres of grass mown, 18; not mown, 10; 25 tons of hay harvested.
- 4 horses, 7 milch cows, 7 other cattle, 7 calves dropped, 1 cattle sold living, 1 cattle slaughtered, 1 cattle died, strayed, stolen and not recovered, 400 lbs. butter, 5 swine, 60 poultry in barnyard, 250 eggs produced.
- 11 acres Indian corn, 300 bu. Indian corn, 10 acres oats, 350 bu. oats, 5 acres rye, 60 bu. rye, 8 acres wheat, 125 bu. wheat, 3/4 acre Irish potatoes, 50 bu. Irish potatoes, 1/16 acre sweet potatoes, 10 bu. sweet potatoes.
- 1 acre apple orchard.
- 15 cords of wood cut valued at $15.

*Federal census lists Isaac, 34; Emeline, 27 (wife); Adam V., 9; Charles V., 7; Henry V., 2.*

## Israel Musser

- Owner
- 60 tilled acres, 8 acres permanent meadows, 5 acres woodland forests, 2 acres unimproved land, valued at $3000.

- Farm machinery valued at $600; livestock, $600; fences, $100; fertilizers, $100; wages for farm labor, $200; weeks of hired labor, 25; estimated value of all farm productions, $700.
- Acres of grass mown, 15; not mown, 10; 25 tons of hay harvested.
- 3 horses, 5 milch cows, 4 other cattle, 5 calves dropped, 1 cattle sold living, 1 cattle slaughtered, 200 lbs. butter, 6 swine, 50 poultry in barnyard, 200 dozen eggs produced.
- 8 acres Indian corn, 250 bu. Indian corn, 8 acres oats, 300 bu. oats, 1 acre rye, 7 bu. rye, 9 acres wheat, 180 bu. wheat, 1 acre Irish potatoes, 100 bu. Irish potatoes.
- 4 acres with 25 apple trees.
- 10 cords of wood cut valued at $10.

*Federal census lists Israel, 29; Catherine, 27 (wife); Lydia A., 5; Peter, 3; Reuben, 1; Sarah, 60 (widowed mother).*

### Moses Horning

- Owner
- 66 tilled acres, 7 acres permanent meadows, 8 acres woodland forests, 17 acres unimproved land, valued at $7000.
- Farm machinery valued at $600; livestock, $700; fences, $10; fertilizers, $70; wages for farm labor, $180; weeks of hired labor, 25; estimated value of all farm productions, $1200.
- Acres of grass mown, 26; not mown, 4; 30 tons hay harvested.
- 4 horses, 8 milch cows, 10 other cattle, 8 calves dropped, 3 cattle sold living, 1 cattle slaughtered, 750 lbs. butter, 5 swine, 50 poultry in barnyard, 250 dozen eggs produced.
- 12 acres Indian corn, 500 bu. Indian corn, 9 acres oats, 300 bu. oats, 5 acres rye, 60 bu. rye, 10 acres wheat, 175 bu. wheat, 1/2 acre Irish potatoes, 30 bu. Irish potatoes, 1/4 acre sweet potatoes, 15 bu. sweet potatoes, 2 acres tobacco, 2600 lbs. tobacco.
- 1 acre with 25 apple trees; total value of orchard products, $5.
- 10 cords of wood cut valued at $20.

*Federal census lists Moses, 50; Lavina, 47 (wife); Franklin G., 18; Mary G., 17; Benjamin G., 14; Israel G., 12; Moses G., 10; Noah G., 5; Ann Huber, 87 (widowed mother).*

### Jacob Musser, Jr.

- Rents for shares of products.
- 47 tilled acres, 6 acres permanent meadows, 11 acres unimproved land, valued at $4000.
- Farm machinery valued at $125; livestock, $300; fences, $25; fertilizers, $10; wages for farm labor, $25; weeks of hired labor, 4; estimated value of all farm productions, $500.
- Acres of grass mown, 15; not mown, 4; 20 tons of hay harvested.
- 3 horses, 6 milch cows, 3 other cattle, 6 calves dropped, 2 cattle sold living, 1 cattle slaughtered, 250 lbs. butter, 4 swine, 60 poultry in barnyard, 250 dozen eggs produced.
- 9 acres Indian corn, 200 bu. Indian corn, 7 acres oats, 220 bu. oats, 3 acres rye, 10 bu. rye, 9 acres wheat, 135 bu. wheat, 1 acre Irish potatoes, 75 bu. Irish potatoes, 1/16 acre sweet potatoes, 4 bu. sweet potatoes.
- 2 acres with 25 apple trees.
- 12 cords of wood cut valued at $18.

*Federal census lists Jacob, Sr., 58; Magdalina, 56; Lizzie, 24; Jacob, Jr., 22; Lydia, 22.*

### James Witwer

- Owner
- 14 tilled acres, 1/2 acre permanent meadows, valued at $2000.
- Farm machinery valued at $200; livestock, $200; fences, $10; fertilizers, $40; wages for farm labor, $50; weeks of hired labor, 10; estimated value of all farm productions, $300.
- Acres of grass mown, 3; not mown, 1; 3 tons of hay harvested.
- 1 horse, 2 milch cows, 1 other cattle, 2 calves dropped, 1 cattle purchased, 2 cattle sold living, 1 cattle slaughtered, 200 lbs. butter, 3 swine, 40 poultry in barnyard, 500 dozen eggs produced.
- 3 acres Indian corn, 75 bu. Indian corn, 2 acres oats, 72 bu. oats, 1/4 acre rye, 5 bu. rye, 3 acres wheat, 67 bu. wheat, 1/2 acre Irish potatoes, 40 bu. Irish potatoes, 1/16 acre sweet potatoes, 3 bu. sweet potatoes, 1/2 acre tobacco, 667 lbs. tobacco.
- Market gardens valued at $5.

*Federal census lists James, 45; Magdalena, 44 (wife); Mary Ellen Nelson, 12 (niece).*

### Christian G. Good

- Owner
- 55 tilled acres, 7 acres permanent meadows, 4 acres woodland forests, 7 acres unimproved land, valued at $4000.
- Farm machinery valued at $600; livestock, $700; fences, $50; fertilizers, $60; wages for farm labor, $75; weeks of hired labor, 13; estimated value of all farm productions, $600.
- Acres of grass mown, 20; not mown, 8; 20 tons hay harvested, 2 bu. clover seed, 1 bu. grass seed.
- 3 horses, 5 milch cows, 5 other cattle, 5 calves dropped, 2 cattle slaughtered; 1 cattle died, strayed, stolen, and not recovered; 500 lbs. butter, 3 swine, 40 poultry in barnyard, 250 dozen eggs produced.

- 7 acres Indian corn, 100 bu. Indian corn, 6 acres oats, 200 bu. oats, 1 acre rye, 8 bu. rye, 9 acres wheat, 150 bu. wheat, 1 acre Irish potatoes, 30 bu. Irish potatoes, 1/2 acre tobacco, 650 lbs. tobacco.
- 1 acre with 20 apple trees, 5 bu. apples, 1/4 acre with 50 peach trees, 4 bu. peaches; total value of orchard products, $6.
- Market gardens valued at $10.
- 8 cords of wood cut valued at $10.

*Federal census lists Christian, 38; Hannah, 39 (wife); Daniel B., 11; Sarah B., 7; Anna B., 5; Emma B., 3.*

## Jacob Weinhold

- Owner
- 17 tilled acres, 4 acres permanent meadows, 1 acre woodland forests, valued at $1450.
- Farm machinery valued at $200; livestock, $150; fences, $50; fertilizers, $40; wages for farm labor, $6; weeks of hired labor, 1; estimated value of all farm productions, $200.
- Acres of grass mown, 6; not mown, 2; 8 tons of hay harvested, 1/2 bu. grass seed.
- 1 horse, 3 milch cows, 2 calves dropped, 2 cattle sold living, 200 lbs. butter, 3 swine, 25 poultry in barnyard, 150 dozen eggs produced.
- 3 acres Indian corn, 105 bu. Indian corn, 3 acres oats, 110 bu. oats, 3 acres wheat, 45 bu. wheat, 1 acre Irish potatoes, 18 bu. Irish potatoes.
- 1 acre with 40 apple trees, 75 bu. apples, total value of orchard products, $30.
- Market gardens valued at $10.

*Federal census lists Jacob, 32; Sarah, 25 (wife); James, 8; William, 6; John, 5.*

## Benjamin Witwer

- Owner
- 40 tilled acres, 3 acres permanent meadows, 18 acres woodland forests, 2 acres unimproved land, valued at $4000.
- Farm machinery valued at $400; livestock, $400; fences, $100; fertilizers, $25; wages for farm labor, $25; weeks of hired labor, 4; estimated value of all farm productions, $600.
- Acres of grass mown, 15; not mown, 2; 12 tons of hay harvested, 6 bu. clover seed.
- 2 horses, 4 milch cows, 7 other cattle, 4 calves dropped, 1 cattle slaughtered, 300 lbs. butter, 4 swine, 60 poultry in barnyard, 300 dozen eggs produced.
- 5 acres Indian corn, 100 bu. Indian corn, 4 acres oats, 125 bu. oats, 1/2 acre rye, 8 bu. rye, 6 acres wheat, 122 bu. wheat, 1 acre Irish potatoes, 150 bu. Irish potatoes, 5 bu. sweet potatoes, 1 acre tobacco, 1400 lbs. tobacco.
- 1 acre with 100 apple trees, 50 bu. apples; total value of orchard products, $25.
- Market gardens valued at $10.
- 15 cords of wood cut valued at $20.

*Federal census lists Benjamin, 38; Maria, 46 (wife); James B., 18; Benjamin B., 17; John B., 16; Maggie B., 14; Sarah B., 11; Harvey B., 9; Noah B., 7; Leaman B., 5.*

## William B. Fox

- Owner
- 60 tilled acres, 11 acres permanent meadows, 60 acres woodland forests, 30 acres unimproved land, valued at $4000.
- Farm machinery valued at $1000; livestock, $800; fences, $50; fertilizers, $100; wages for farm labor, $100; weeks of hired labor, 20; estimated value of all farm productions, $1000.
- Acres of grass mown, 32; 20 tons of hay harvested, 2 1/2 bu. clover seed.
- 6 horses, 8 milch cows, 8 other cattle, 7 calves dropped, 1 animal slaughtered; 1 cattle died, strayed, stolen, and not recovered; 400 lbs. butter, 12 swine, 25 poultry in barnyard, 400 dozen eggs produced.
- 13 acres Indian corn, 300 bu. Indian corn, 15 acres oats, 400 bu. oats, 2 acres rye, 24 bu. rye, 11 acres wheat, 200 bu. wheat, 1 1/2 acres Irish potatoes, 100 bu. Irish potatoes, 2 acres tobacco, 2100 lbs. tobacco.
- 3 acres with 50 apple trees.
- Market gardens valued at $5.
- 20 cords of wood cut valued at $20.

*Federal census lists William, 44; Annette, 44 (wife); Milton, 16; Elizabeth, 14; Aaron, 12; Sarah, 10; Emma, 8; Wayne, 6; William, 4; Maggie, 2; Samuel, 80 (father).*

## Daniel Kieffer

- Owner
- 6 tilled acres, 2 acres permanent meadows, 6 acres woodland forests, valued at $1000.
- Farm machinery valued at $50; livestock, $75; fences, $10; fertilizers, $10; estimated value of all farm productions, $150.
- Acres of grass mown, 2; 2 tons of hay harvested.
- 1 horse, 2 milch cows, 1 calf dropped, 1 cattle sold living, 150 lbs. butter, 2 swine, 10 poultry in barnyard, 100 dozen eggs produced.

- 1 acre Indian corn, 20 bu. Indian corn, 1 acre wheat, 25 bu. wheat, 1/4 acre Irish potatoes, 10 bu. Irish potatoes.
- 1/2 acre with 10 apple trees.
- Market gardens valued at $1.
- 6 cords of wood cut valued at $6.

*Federal census lists Daniel, 53; Mary, 54 (wife); Franklin, 18, fencemaker.*

### Daniel Glass (deceased)

- Owner
- 32 tilled acres, 4 acres permanent meadows, 119 acres woodland forests, valued at $1500.
- Farm machinery valued at $200; livestock, $175; fences, $30; fertilizers, $20; wages paid for farm labor, $40; weeks of hired labor, 6; estimated value of all farm productions, $600.
- Acres of grass mown, 10; 4 tons of hay harvested, 2 1/2 bu. clover seed.
- 2 horses, 3 milch cows, 4 other cattle, 5 calves dropped, 300 lbs. butter, 2 swine, 25 poultry in barnyard, 400 dozen eggs produced.
- 3 acres Indian corn, 100 bu. Indian corn, 4 acres oats, 90 bu. oats, 4 acres wheat, 52 bu. wheat, 1 acre Irish potatoes, 25 bu. Irish potatoes, 1 acre tobacco, 1500 lbs. tobacco.
- 3 acres with 75 apple trees, 15 bu. apples, 2 acres with 200 peach trees, 30 bu. peaches; total value of orchard products, $15.
- Market gardens valued at $75.
- 6 cords of wood cut valued at $7.

*Federal census lists Lydia, 47 (widow); Lydia A., 22; Catharine, 20; Emma, 17; Barton, 14; Frank, 11; Sarah A., 7; Mary E., 4; Abraham Brendel, 73, and Susan Brendel, 68 (Lydia's parents).*

### Jacob Horst

- Owner
- 25 tilled acres, 3 acres permanent meadows, 5 acres woodland forests, 2 acres unimproved land, valued at $2000.
- Farm machinery valued at $200; livestock, $200; fences, $30; fertilizers, $20; wages paid for farm labor, $20; weeks of hired labor, 4; estimated value of all farm productions, $200.
- Acres of grass mown, 4; not mown, 2; 5 tons of hay harvested, 1 bu. clover seed.
- 2 horses, 4 milch cows, 3 other cattle, 4 calves dropped; 11 cattle died, strayed, stolen, and not recovered; 250 lbs. butter, 5 swine, 50 poultry in barnyard, 250 dozen eggs produced.
- 4 acres Indian corn, 100 bu. Indian corn, 4 acres oats, 90 bu. oats, 4 acres wheat, 90 bu. wheat, 1/2 acre Irish potatoes, 20 bu. Irish potatoes, 1/2 acre tobacco, 700 lbs. tobacco.
- 1/2 acre with 10 apple trees, 2 bu. apples.
- 6 cords of wood cut valued at $6.

*Federal census lists Jacob, 44; Rebecca, 43 (wife); Barton, 18; Jacob, 15; David, 13; Abraham, 12; John, 9; Martin, 4; Judith, 79 (mother).*

### Isaac Keiffer

- Owner
- 32 tilled acres, 4 acres permanent meadows, 10 acres woodland forests, 5 acres unimproved land, valued at $1800.
- Farm machinery valued at $100; livestock, $150; fences, $50; fertilizers, $5; weeks of hired labor, 5; estimated value of all farm productions, $200.
- Acres of grass mown, 12; not mown, 10; 16 tons of hay harvested, 2 bu. clover seed.
- 1 horse, 4 milch cows, 2 other cattle, 3 calves dropped, 5 cattle sold living, 400 lbs. butter, 4 swine, 40 poultry in barnyard, 250 dozen eggs produced.
- 4 acres Indian corn, 75 bu. Indian corn, 4 acres oats, 110 bu. oats, 1/2 acre rye, 11 bu. rye, 4 acres wheat, 76 bu. wheat, 1/2 acre Irish potatoes, 40 bu. Irish potatoes.
- 1/2 acre with 20 apple trees, 6 bu. apples; total value of orchard products, $5.
- Market gardens valued at $30.
- 5 cords of wood cut valued at $5.

*Federal census lists Isaac, 40; Caroline, 38 (wife); Barton, 20; Amanda, 18; Elizabeth, 16; Catharine, 13; William, 11; Charles, 9; Henry, 7; Ellen, 5; Emma, 3; Caroline, 1.*

### John Fox

- Owner
- 30 tilled acres, 2 acres permanent meadows, 11 acres woodland forests, 2 acres unimproved land, valued at $1100.
- Farm machinery valued at $200; livestock, $200; fences, $30; fertilizers, $8; estimated value of all farm productions, $100.
- Acres of grass mown, 4; not mown, 4; 4 tons of hay harvested.
- 1 horse, 3 milch cows, 2 other cattle, 3 calves dropped, 3 cattle sold living, 100 lbs. butter, 3 swine, 12 poultry in barnyard, 24 dozen eggs produced.

- 3 acres Indian corn, 20 bu. Indian corn, 2 1/2 acres oats, 73 bu. oats, 4 acres wheat, 45 bu. wheat, 1/2 acre Irish potatoes, 20 bu. Irish potatoes.
- 1/2 acre with 10 apple trees, 5 bu. apples; total value of orchard products, $4.
- Market gardens valued at $5.
- 11 cords of wood cut valued at $10.

*Federal census lists John, 57, a carpenter; Catharine, 50 (wife); Henderson, 16, a farm laborer; Thomas, 19, a farm laborer; Ellen, 2.*

## Isaac Griffy

- Owner
- 12 tilled acres, 1 acre permanent meadows, 1 acre woodland forests, valued at $700.
- Farm machinery valued at $50; livestock, $25; fences, $10; fertilizers, $10; wages paid for farm labor, $10; weeks of hired labor, 2; estimated value of all farm productions, $100.
- Acres of grass mown, 2; 2 tons of hay harvested.
- 1 horse, 1 swine, 10 poultry in barnyard, 20 dozen eggs produced.
- 1 acre Indian corn, 15 bu. Indian corn, 1/2 acre rye, 7 bu. rye, 2 acres wheat, 19 bu. wheat, 1/4 acre Irish potatoes, 10 bu. Irish potatoes.
- 1/2 acre with 30 apple trees, 2 bu. apples; total value of orchard products, $1.
- Market gardens valued at $1.
- 11 cords of wood cut valued at $10.

*Federal census lists Isaac, 63; Betsey, 65 (wife).*

## Isaac Burkhart

- Owner
- 70 tilled acres, 8 acres permanent meadows, 5 acres woodland forests, 35 acres unimproved land, valued at $6000.
- Farm machinery valued at $1000, livestock, $1000, fences, $100, fertilizers, $75; wages paid for farm labor, $100; weeks of hired labor, 20; estimated value of all farm productions, $1200.
- Acres of grass mown, 50; not mown, 15; 40 tons of hay harvested, 4 bu. clover seed, 2 bu. clover seed.
- 3 horses, 6 milch cows, 5 other cattle, 6 calves dropped, 8 cattle purchased, 6 cattle sold, 2 cattle slaughtered; 7 cattle died, strayed, stolen, and not recovered; 1000 lbs. butter, 9 swine, 50 poultry in barnyard, 4 other poultry, 250 dozen eggs produced.
- 4 acres Indian corn, 300 bu. Indian corn, 10 acres oats, 600 bu. oats, 1/2 acre rye, 10 bu. rye, 15 acres wheat, 250 bu. wheat, 1 acre Irish potatoes, 30 bu. Irish potatoes, 1/2 acre sweet potatoes, 20 bu. sweet potatoes.
- 2 acres with 30 apple trees.
- Market gardens valued at $20.
- 10 cords of wood cut valued at $12.

*Federal census lists Isaac, 60; Esther, 52 (wife); David C., 21; Amos W., 13; Mary W., 11; Rebecca, 10.*

## Isaac G. Shupp

- Owner
- 25 tilled acres, 3 acres permanent meadows, 25 acres woodland forests, 20 acres unimproved land, valued at $2500.
- Farm machinery valued at $200; livestock, $100; fences, $100; fertilizers, $50; wages paid for farm labor, $100; weeks of hired labor, 20; estimated value of all farm productions, $500.
- Acres of grass mown, 9; 15 tons of hay harvested; 2 bu. clover seed.
- 2 mules and asses, 6 milch cows, 2 other cattle, 1 calf dropped, 1 cattle sold living, 150 lbs. butter, 5 swine, 25 poultry in barnyard, 100 dozen eggs produced.
- 1/2 acre buckwheat, 4 bu. buckwheat, 5 acres Indian corn, 100 bu. Indian corn, 5 bu. oats, 200 bu. oats, 1/4 acre rye, 4 bu. rye, 5 acres wheat, 100 bu. wheat, 1/2 acre Irish potatoes, 10 bu. Irish potatoes.
- 1/2 acre with 10 apple trees.
- Market gardens valued at $1.
- 10 cords of wood cut valued at $10.

*Federal census lists Isaac, 45; Barbara, 38 (wife); Nathan H., 16; Barton H., 15; Emma H., 12; Henry H., 11; Wayne H., 9; Clara H., 7; David H., 2; Mary H., 1.*

## Christian Miller

- Owner
- 30 tilled acres, 4 acres permanent meadows, 7 acres woodland forests, 10 acres unimproved land, valued at $3000.
- Farm machinery valued at $100; livestock, $200; fences, $10; fertilizers, $20; wages paid for farm labor, $25; weeks of hired labor, 4; estimated value of all farm productions, $500.
- Acres of grass mown, 20; not mown, 10; 20 tons of hay harvested.
- 2 horses, 6 milch cows, 1 other cattle, 6 calves dropped, 2 cattle purchased, 2 cattle sold living, 400 lbs. butter, 4 swine, 40 poultry in barnyard, 80 dozen eggs produced.

- 4 acres Indian corn, 100 bu. Indian corn, 4 acres oats, 160 bu. oats, 7 acres wheat, 130 bu. wheat, 1/4 acre Irish potatoes, 60 bu. Irish potatoes.
- 1/2 acre with 18 apple trees.
- 6 cords of wood cut valued at $6.

*Federal census lists Christian, 53; Susan, 55 (wife); Henry 25, a carpenter.*

### David Fox

- Owner
- 27 tilled acres, 6 acres permanent meadows, 1 acre woodland forests, 7 acres unimproved land, valued at $1600.
- Livestock valued at $200; fences, $100; fertilizers, $100; wages paid for farm labor, $400; weeks of hired labor, 52; estimated value of all farm productions, $400.
- Acres of grass mown, 13; 8 tons of hay harvested, 4 bu. clover seed.
- 1 horse, 4 milch cows, 3 calves dropped, 1 cattle purchased, 1 cattle sold living, 2 cattle died, strayed, stolen and not recovered, 300 lbs. butter, 6 swine, 30 poultry in barnyard, 150 dozen eggs produced.
- 4 acres Indian corn, 150 bu. Indian corn, 4 acres oats, 125 bu. oats, 1/2 acre rye, 17 bu. rye, 4 acres wheat, 81 bu. wheat, 1/2 acre Irish potatoes, 50 bu. Irish potatoes, 1/12 acre sweet potatoes, 3 bu. sweet potatoes, 3 acres tobacco, 150 lbs. tobacco.
- 1 acre with 25 apple trees; total value of orchard products, $1.
- Market gardens valued at $21.

*Federal census lists David, 45; Rebecca, 43 (wife); Samuel, 21; Frank, 19; John, 18; Henry, 17; Elizabeth, 14; Adaline, 9; Wayne Kline, 10 (foundling).*

### Jeremiah Brendle

- Owner
- 18 tilled acres, 4 acres permanent meadows, 1 acre woodland forests, 5 acres unimproved land, valued at $900.
- Farm machinery valued at $200; livestock, $100; fences, $75; fertilizers, $40; wages paid for farm labor, $10; weeks of hired labor, 2; estimated value of all farm productions, $200.
- Acres of grass mown, 5; 4 tons of hay harvested.
- 1 horse, 2 milch cows, 1 calf dropped, 2 cattle died, strayed, stolen and not recovered, 150 lbs. butter, 1 swine, 30 poultry in barnyard, 180 dozen eggs produced.
- 3 acres Indian corn, 70 bu. Indian corn, 2 acres oats, 50 bu. oats, 2 acres wheat, 25 bu. wheat, 1 1/2 acres Irish potatoes, 100 bu. Irish potatoes.
- 1/2 acre with 30 apple trees.
- Market gardens valued at $20.
- 4 cords of wood cut valued at $4.

*Federal census lists Jeremiah, 35; Susan, 28 (wife); Margaret, 7; Caroline, 3.*

### John Steffe

- Rents for fixed money rental.
- 45 tilled acres, 10 acres permanent meadows, 40 acres woodland forests, 25 acres unimproved land, valued at $3000.
- Farm machinery valued at $500; livestock, $200; fences, $100; fertilizers $50; wages paid for farm labor, $100; weeks of hired labor, 20; estimated value of all farm productions, $600.
- Acres of grass mown, 25; not mown, 30; 25 tons of hay harvested.
- 4 horses, 5 milch cows, 3 other cattle, 3 calves dropped, 300 lbs. butter, 1 swine, 25 poultry in barnyard, 100 dozen eggs produced.
- 9 acres Indian corn, 300 bu. Indian corn, 5 acres oats, 100 bu. oats, 1 acre rye, 10 bu. rye, 12 acres wheat, 130 bu. wheat, 3/4 acre Irish potatoes, 25 bu. Irish potatoes.
- 1/2 acre with 30 apple trees.
- Market gardens valued at $5.

*Federal census lists John, 41; Lydia, 38 (wife); Noah, 18; Elizabeth, 15; Anna, 12; Agnes, 10, Caroline, 7; John, 5; Catherine, 4; Mahala, 1; Elam, 3/12 (Feb.).*

### Milton Kern

- Rents for shares of products.
- 60 tilled acres, 10 acres permanent meadows, 4 acres woodland forests, 15 acres unimproved land, valued at $4000.
- Farm machinery valued at $800; livestock, $400; fences, $100; fertilizers, 50; wages paid for farm labor, $75; weeks of hired labor, 12; estimated value of all farm productions, $600.
- Acres of grass mown, 30; not mown, 25; 25 tons of hay harvested; 3 bu. clover seed; 1/2 bu. grass seed.
- 3 horses, 4 milch cows, 4 other cattle, 4 calves dropped, 1 cattle died, strayed, stolen, and not recovered, 300 lbs. butter, 5 swine, 15 poultry in barnyard, 50 dozen eggs produced.
- 53 acres Indian corn, 300 bu. Indian corn, 14 acres oats, 300 bu. oats, 14 acres wheat, 150 bu. wheat, 1 acre Irish potatoes, 20 bu. Irish potatoes.

- 3 acres with 25 apple trees; 5 bu. apples; total value of orchard products, $2.
- Market gardens valued at $5.
- 5 cords of wood cut valued at $5.

*Federal census lists Milton, 22; Mary, 20 (wife); William, 1/3 (Jan.); Charles Clark, 15 (sister's son); Isaac Kern, 70 (widowed father); Elizabeth, 34 (sister); Hattie, 26 (sister).*

## Jacob Fritz

- Owner
- 25 tilled acres, 2 acres permanent meadows, 2 acres unimproved land, valued at $950.
- Farm machinery valued at $150; livestock, $200; fences, $35; fertilizers, $35; wages paid for farm labor, $15; weeks of hired labor, 2; estimated value of all farm productions, $300.
- Acres of grass mown, 10; 10 tons of hay harvested; 2 bu. clover seed.
- 3 milch cows, 1 other cattle, 2 calves dropped, 2 cattle purchased, 200 lbs. butter, 4 swine, 20 poultry in barnyard, 140 dozen eggs produced.
- 4 acres Indian corn, 70 bu. Indian corn, 3 acres oats, 70 bu. oats, 1 acre rye, 4 bu. rye, 2 acres wheat, 10 bu. wheat, 1 1/2 acres Irish potatoes, 70 bu. Irish potatoes.
- 1 acre with 20 apple trees.

*Federal census lists Jacob, 27; Susan, 29 (wife); Sarah E., 6.*

## John M. Good

- Owner
- 40 tilled acres, 5 acres permanent meadows, 2 acres woodland forests, 2 acres improved land, valued at $3000.
- Farm machinery valued at $300; livestock, $400; fences, $50; fertilizers, $50; wages paid for farm labor, $50; weeks of hired labor, 15; estimated value of all farm productions, $680.
- Acres of grass mown, 10; not mown, 5; 20 tons of hay harvested, 2 bu. clover seed.
- 4 horses, 5 milch cows, 4 other cattle, 5 calves dropped, 1 cattle slaughtered, 250 lbs. butter, 4 swine, 30 poultry in barnyard, 120 dozen eggs produced.
- 5 acres Indian corn, 150 bu. Indian corn, 5 acres oats, 300 bu. oats, 1 acre rye, 5 bu. rye, 4 acres wheat, 80 bu. wheat, 1 bu. Irish potatoes, 20 bu. Irish potatoes, 2 acres tobacco, 1800 lbs. tobacco.
- 2 acres with 10 apple trees.
- 5 cords of wood cut valued at $5.

*Federal census lists John, 65; Lydia, 60 (wife); Allison, 31, son.*

## Isaac M. Good

- Owner
- 30 tilled acres, 12 acres permanent meadows, 2 acres woodland forests, 2 acres unimproved land valued at $2000.
- Farm machinery valued at $300; livestock, $400; fences, $50; fertilizers, $50; wages paid for farm labor, $50; weeks of hired labor, 15; estimated value of all farm productions, $400.
- Acres of grass mown, 10; not mown, 5; 20 tons of hay harvested, 2 bu. clover seed.
- 4 horses, 6 milch cows, 5 other cattle, 5 calves dropped, 2 cattle sold living, 1 cattle slaughtered, 1 cattle died, strayed, stolen, and not recovered, 250 lbs. butter, 7 swine, 20 poultry in barnyard, 120 dozen eggs produced.
- 6 acres Indian corn, 150 bu. Indian corn, 5 acres oats, 200 bu. oats, 1 acre rye, 10 bu. rye, 5 acres wheat, 100 bu. wheat, 1/2 acre Irish potatoes, 10 bu. Irish potatoes.
- 1 acre with 10 apple trees.
- Market gardens valued at $5.
- 10 cords wood cut valued at $10.

*Federal census lists Isaac, 48; Louisa Ann, 37 (wife); Alvester S., 11; Calvin S., 9; Lizzie S., 5; Harvey S., 1.*

## David Good

- Owner
- 40 tilled acres, 5 acres permanent meadows, 10 acres woodland forests, 15 acres unimproved land, valued at $4000.
- Farm machinery valued at $200; livestock, $250; fences, $100; fertilizers, $50; wages paid for farm labor, $25; weeks of hired labor, 10; estimated value of all farm productions, $400.
- Acres of grass mown, 12; not mown, 10; 20 tons of hay harvested, 2 bu. clover seed, 1/2 bu. grass seed.
- 2 horses, 4 milch cows, 2 other cattle, 4 calves dropped, 1 cattle purchased, 1 cattle sold living, 1 cattle slaughtered, 1 cattle died, strayed, stolen and not recovered, 200 lbs. butter, 2 swine, 15 poultry in barnyard, 11 other poultry, 150 dozen eggs produced.
- 2 acres Indian corn, 60 bu. Indian corn, 3 acres oats, 75 bu. oats, 1 acre rye, 1 bu. rye, 2 acres wheat, 33 bu. wheat, 1/2 acre Irish potatoes, 25 bu. Irish potatoes.
- 1 acre with 30 apple trees, 1 bu. apples.
- 40 cords of wood cut valued at $60.

*Federal census lists David, 49 (widower); Lydia, 59 (cousin).*

### Mariah Shiffer

- Owner
- 56 tilled acres, 5 acres permanent meadows, 5 acres woodland forests, 7 acres unimproved land, valued at $6000.
- Farm machinery valued at $500; livestock, $900; fences, $100; fertilizers, $150; wages paid for farm labor, $50; weeks of hired labor, 10; estimated value of all farm productions, $900.
- Acres of grass mown, 15; not mown, 5; 36 tons of hay harvested, 2 bu. clover seed, 1 bu. grass seed.
- 4 horses, 8 milch cows, 6 other cattle, 8 calves dropped, 1 cattle sold living, 1 cattle slaughtered, 700 lbs. butter, 6 swine, 50 poultry in barnyard, 400 dozen eggs produced.
- 10 acres Indian corn, 200 bu. Indian corn, 10 acres oats, 300 bu. oats, 1 acre rye, 5 bu. rye, 13 acres wheat, 180 bu. wheat, 1/2 acre Irish potatoes, 50 bu. Irish potatoes.
- 1 acre with 25 apple trees, 1 bu. apples; total value of orchard products, $1.
- 10 cords of wood cut valued at $10.

*Federal census lists Mariah, 58 (widow); Israel, 26; Martin, 24; Isaac, 22; Richard, 20; Susan, 19; John, 17; Abraham, 13; Jacob, 11.*

### Hannah Eshleman

- Owner
- 70 tilled acres, 10 acres permanent meadows, 30 acres woodland forests, 51 acres unimproved land, valued at $5000.
- Farm machinery valued at $700; livestock, $600; fences, $300; fertilizers, $30; wages paid for farm labor, $25; weeks of hired labor, 10; estimated value of all farm productions, $700.
- Acres of grass mown, 20; not mown, 4; 30 tons of hay harvested, 3 bu. clover seed, 5 bu. grass seed.
- 4 horses, 5 milch cows, 7 other cattle, 5 calves dropped, 1 cattle sold living, 250 lbs. butter, 10 swine, 60 poultry in barnyard, 300 dozen eggs produced.
- 6 acres Indian corn, 200 bu. Indian corn, 6 acres oats, 160 bu. oats, 1/4 acre rye, 2 bu. rye, 9 acres wheat, 120 bu. wheat, 1/2 acre Irish potatoes, 25 bu. Irish potatoes, 1/2 acre tobacco, 300 lbs. tobacco.
- 2 acres with 25 apple trees.
- Market gardens valued at $5.
- 3 cords of wood cut valued at $3.

*Federal census lists Hannah, 65 (widow); Isaac, 22; Absalom, 17; Rose A., 13; Samuel, 39, farm laborer; Louise, 35 (wife).*

### Samuel Weinhold

- Owner
- 50 tilled acres, 10 acres permanent meadows, 4 acres woodland forests, 11 acres unimproved land, valued at $4000.
- Farm machinery valued at $600; livestock, $600; fences, $100; fertilizers, $50; wages paid for farm labor, $10; weeks of hired labor, 10; estimated value of all farm productions, $400.
- Acres of grass mown, 20; not mown, 10; 30 tons of hay harvested, 2 bu. clover seed, 7 bu. grass seed.
- 3 horses, 1 mules and asses, 4 milch cows, 4 other cattle, 4 calves dropped, 2 cattle purchased, 3 cattle sold living, 1 cattle slaughtered, 1 cattle died, strayed, stolen and not recovered, 250 lbs. butter, 4 swine, 60 poultry in barnyard, 300 dozen eggs produced.
- 6 acres Indian corn, 200 bu. Indian corn, 7 acres oats, 300 bu. oats, 1 acre rye, 15 bu. rye, 12 acres wheat, 250 bu. wheat, 1/2 acre Irish potatoes, 15 bu. Irish potatoes.
- 1 acre with 20 apples, 10 bu. apples; total value of orchard products, $5.
- 5 cords of wood cut valued at $5.

*Federal census lists Samuel, 31; Lizzie, 28 (wife); Aliza, 5; John, 3; Lydia, 2; William, 6/12 (Dec.).*

### Moses Huyard

- Owner
- 5 tilled acres, valued at $1220
- Farm machinery valued at $25; livestock, $100; fences, $25; fertilizers, $25; wages paid for farm labor, $25; weeks of hired labor, 4; estimated value of all farm productions, $250.
- Acres of grass mown, 2, 2 tons of hay harvested.
- 1 milch cow, 1 calf dropped, 1 cattle sold living, 100 lbs. butter, 4 swine, 30 poultry in barnyard, 200 dozen eggs produced.
- 1 acre Indian corn, 25 bu. Indian corn, 2 acres wheat, 20 bu. wheat, 1/2 acre Irish potatoes, 20 bu. Irish potatoes.

*Federal census lists Moses, 38, a carpenter; Lydia, 37 (wife); Louisa, 17; Isaac, 15, farm laborer; Sara A., 12; William, 9; Amilia, 6.*

### Amos Schneder

- Owner
- 29 tilled acres, 1 acre permanent meadows, 3 acres woodland forests, 1 acre unimproved land, valued at $1600.

- Farm machinery valued at $150; livestock, $300; fences, $10; fertilizers, $25; wages paid for farm labor, $30; weeks of hired labor, 4; estimated value of all farm productions, $400.
- Acres of grass mown, 10; not mown, 2; 20 tons of hay harvested, 3 bu. clover seed.
- 2 horses, 3 milch cows, 4 other cattle, 3 cattle sold living, 1 cattle slaughtered, 1 cattle died, strayed, stolen and not recovered, 300 lbs. butter, 6 sheep on hand, 6 lambs dropped, 3 sheep sold living, 1 sheep died of disease, 6 fleeces, 10 lbs. wool, 2 swine, 50 poultry in barnyard, 200 dozen eggs produced.
- 3 acres Indian corn, 150 bu. Indian corn, 2 acres oats, 90 bu. oats, 1/4 acre rye, 7 bu. rye, 3 acres wheat, 55 bu. wheat, 1/4 acre Irish potatoes, 10 bu. Irish potatoes.
- 1/2 acre with 20 apple trees, 3 bu. apples; total value of orchard products, $2.
- Market gardens valued at $2.
- 4 cords of wood cut valued at $4.

*Federal census lists Amos, 73; Catherine, 72 (wife).*

### John Geigley

- Owner
- 54 tilled acres, 6 acres permanent meadows, 7 acres unimproved land, valued at $5000.
- Farm machinery valued at $600; livestock, $375; fences, $50; fertilizers, $100; wages paid for farm labor, $325; weeks of hired labor, 58; estimated value of all farm productions, $1000.
- Acres of grass mown, 21; 25 tons of hay harvested.
- 3 horses, 5 milch cows, 3 other cattle, 4 calves dropped, 2 cattle purchased, 3 cattle sold living, 1 cattle slaughtered, 350 lbs. butter, 10 swine, 50 poultry in barnyard, 5 other poultry, 520 dozen eggs produced.
- 7 acres Indian corn, 300 bu. Indian corn, 6 acres oats, 300 bu. oats, 1/4 acre rye, 6 bu. rye, 11 acres wheat, 222 bu. wheat, 1/2 acre Irish potatoes, 25 bu. Irish potatoes, 1/16 acre sweet potatoes, 20 bu. sweet potatoes, 1 acre tobacco, 1760 lbs. tobacco.
- 1/4 acre with 23 apple trees, 1 bu. apples; 1/2 acre with 32 peach trees, 6 bu. peaches; total value of orchard products, $5.

*Federal census lists John, 59; Matilda, 59 (wife); Mary A., 19; Emma, 15; Frank Carpenter, 15 (nephew).*

### Amos Goshen, Dec'd.

- Owner
- 16 tilled acres, 2 acres unimproved land, valued at $2000.
- Farm machinery valued at $100; livestock, $250; fences, $50; fertilizers, $25; wages paid for farm labor, $20; weeks of hired labor, 2; estimated value of all farm productions, $600.
- Acres of grass mown, 6; 10 tons of hay harvested.
- 1 horse, 2 milch cows, 2 calves dropped, 1 cattle purchased, 2 cattle sold living, 1 cattle slaughtered, 300 lbs. butter, 2 swine, 40 poultry in barnyard, 400 dozen eggs produced.
- 3 acres Indian corn, 125 bu. Indian corn, 3 acres oats, 100 bu. oats, 1/2 acre rye, 6 bu. rye, 3 acres wheat, 70 bu. wheat, 1/4 acre Irish potatoes, 25 bu. Irish potatoes.
- 1/4 acre with 15 apple trees, 5 bu. apples; 1/2 acre with 10 peach trees, 3 bu. peaches; total value of orchard products, $5.

*Federal census lists Hattie, 42 (widow); John, 16; George, 14; Amos, 23; Sarah, 24 (wife); Ellman, 3.*

### Jacob Schnader

- Owner
- 60 tilled acres, 10 acres permanent meadows, 10 acres unimproved land, valued at $6000.
- Farm machinery valued at $500; livestock, $500; fences, $100; fertilizers, $100; wages paid for farm labor, $100; weeks of hired labor, 20; estimated value of all farm productions, $1700.
- Acres of grass mown, 15; not mown, 10; 20 tons of hay harvested, 2 bu. clover seed, 2 bu. grass seed.
- 5 horses, 9 milch cows, 9 other cattle, 9 calves dropped, 1 cattle purchased, 2 cattle sold living, 1 cattle slaughtered, 1 cattle died, strayed, stolen and not recovered, 600 lbs. butter, 7 swine, 50 poultry in barnyard, 400 dozen eggs produced.
- 15 acres Indian corn, 300 bu. Indian corn, 15 acres oats, 300 bu. oats, 2 acres rye, 20 bu. rye, 15 acres wheat, 300 bu. wheat, 1 acre Irish potatoes, 50 bu. Irish potatoes, 1 acre tobacco, 1500 lbs. tobacco.
- 1/2 acre with 10 apple trees, 2 bu. apples; 1/2 acre with 10 peach trees, 2 bu. peaches; total value of orchard products, $5.
- Market gardens valued at $20.
- 10 cords of wood cut valued at $10.

*Federal census lists Jacob, 40; Marie, 39 (wife); Emma, 19; Irwin, 15; Polly Mose, servant, 41; James Spotts, servant, 10.*

### Martin B. Good

- Rents for shares of products.
- 62 tilled acres, 16 permanent meadows, 35 acres woodland forests, 10 acres unimproved land, valued at $8000.
- Farm machinery valued at $600; livestock, $300; fences, $75; fertilizers, $100; wages paid for farm labor, $150; weeks of hired labor, 25; estimated value of all farm productions, $910.
- Acres of grass mown, 25; not mown, 3; 25 tons of hay harvested, 5 bu. clover seed, 5 bu. grass seed.

- 4 horses, 5 milch cows, 7 other cattle, 5 calves dropped, 6 cattle purchased, 4 cattle sold living, 1 cattle slaughtered, 1 cattle died, strayed, stolen and not recovered, 400 lbs. butter, 2 sheep on hand, 3 lambs dropped, 5 sheep sold living, 2 fleeces, 5 lbs. wool, 10 swine, 50 poultry in barnyard, 300 dozen eggs produced.
- 11 acres Indian corn, 300 bu. Indian corn, 12 acres oats, 260 bu. oats, 12 acres wheat, 230 bu. wheat, 1/2 acre Irish potatoes, 50 bu. Irish potatoes.
- 4 acres with 50 apple trees.
- 8 cords of wood cut valued at $10.

*Federal census lists Martin B., 30; Catherine, 29 (wife); Mahlon M., 4; Anna L., 2; John H., 66, retired farmer; Lavinia, 55 (wife); Hattie A., 33; Israel, 18, a teacher; Sarah, 49 (John's sister).*

## Jacob Renninger

- Owner
- 30 tilled acres, 2 acres permanent meadows, 3 acres unimproved land valued at $4000.
- Farm machinery valued at $100; livestock, $300; fences, $10; fertilizers, $20; wages paid for farm labor, $50; weeks of hired labor, 8; estimated value of all farm productions, $400.
- Acres of grass mown, 12; not mown, 3; 10 tons of hay harvested, 3 bu. clover seed.
- 2 horses, 4 milch cows, 1 other cattle, 3 calves dropped, 1 cattle purchased, 1 cattle slaughtered, 1 cattle died, strayed, stolen and not recovered, 200 lbs. butter, 4 swine, 40 poultry in barnyard, 100 dozen eggs produced.
- 5 acres Indian corn, 200 bu. Indian corn, 5 acres oats, 175 bu. oats, 1 acre rye, 6 bu. rye, 5 acres wheat, 133 bu. wheat, 1/4 acre Irish potatoes, 25 bu. Irish potatoes, 1/32 acre sweet potatoes, 3 bu. sweet potatoes.
- 1 acre with 50 apple trees, 1/2 bu. apples; total value of orchard products, $5.
- 10 pounds of honey.
- Value of forest products, $3.

*Federal census lists Jacob, 69, a blacksmith; Catherine, 64 (wife); Ida, 6 (grandchild).*

## Rudolf Brossman

- Owner
- 64 tilled acres, 3 acres permanent meadows, 300 acres woodland forests, 1 acre unimproved land valued at $6500.
- Farm machinery valued at $300; livestock, $300; fences, $55; fertilizers, $60; wages paid for farm labor, $50; weeks of hired labor, 8; estimated value of all farm productions, $1000.
- Acres of grass mown, 20; not mown, 3; 25 tons of hay harvested.
- 2 horses, 4 milch cows, 4 other cattle, 4 calves dropped, 1 cattle purchased, 1 cattle sold living, 1 cattle slaughtered, 500 lbs. butter, 5 swine, 50 poultry in barnyard, 500 dozen eggs produced.
- 3 acres Indian corn, 100 bu. Indian corn, 2 acres oats, 100 bu. oats, 4 acres wheat, 100 bu. wheat, 1/4 acre Irish potatoes, 25 bu. Irish potatoes, 1/32 acre sweet potatoes, 3 bu. sweet potatoes.
- 2 acres with 50 apple trees, 1/32 acre with 5 peach trees; total value of orchard products, $5.
- 10 cords of wood cut valued at $15.

*Federal census lists Rudolf, 40; Elizabeth, 37 (wife); John Brossman, 74, retired farmer; Hannah, 78 (wife).*

## Solomon Good

- Owner
- 49 tilled acres, 5 acres permanent meadows, 19 acres woodland forests, 2 acres unimproved land, valued at $4500.
- Farm machinery valued at $300; livestock, $500; fences, $25; fertilizers, $40; wages paid for farm labor, $10; weeks of hired labor, 2; estimated value of all farm productions, $800.
- Acres of grass mown, 15; not mown, 2; 20 tons of hay harvested, 2 bu. clover seed.
- 6 milch cows, 3 other cattle, 6 calves dropped, 5 cattle purchased, 8 cattle sold living, 1 cattle slaughtered, 1 cattle died, strayed, stolen and not recovered, 500 lbs. butter, 2 swine, 25 poultry in barnyard, 300 dozen eggs produced.
- 9 acres Indian corn, 200 bu. Indian corn, 6 acres oats, 349 bu. oats, 1 acre rye, 10 bu. rye, 10 acres wheat, 215 bu. wheat, 1/2 acre Irish potatoes, 50 bu. Irish potatoes, 1/32 acre sweet potatoes, 8 bu. sweet potatoes, 3/4 acres tobacco, 700 bu. tobacco.
- 1 acre with 25 apple trees.

*Federal census lists Solomon, 49; Fanny, 41 (wife); Noah G., 17; Aaron G., 16; Peter G., 12; William G., 2.*

## David Bender

- Owner
- 70 tilled acres, 2 acres permanent meadows, 7 acres woodland forests, valued at $7000.
- Farm machinery valued at $1000; livestock, $800; fences, $50; fertilizers, $90; wages paid for farm labor, $100; weeks of hired labor, 40; estimated value of all farm productions, $1000.
- Acres of grass mown, 25; not mown, 8; 40 tons of hay harvested.

- 3 horses, 2 mules and asses, 5 milch cows, 6 other cattle, 5 calves dropped, 1 cattle purchased, 3 cattle sold living, 1 cattle slaughtered, 600 lbs. butter, 3 swine, 80 poultry in barnyard, 800 dozen eggs produced.
- 10 acres Indian corn, 500 bu. Indian corn, 10 acres oats, 400 bu. oats, 1 acre rye, 20 bu. rye, 10 acres wheat, 200 bu. wheat, 1/2 acre Irish potatoes, 30 bu. potatoes, 1/32 acre sweet potatoes, 5 bu. sweet potatoes.
- 1 acre with 20 apple trees, 10 bu. apples; total value of orchard products, $3.
- Market gardens valued at $3.

*Federal census lists David, 48; Eliza, 48 (wife); Maria, 22; Caty, 19; David, 17; Eliza, 11; Judith, 76 (aunt).*

## William B. Renninger

- Owner
- 50 tilled acres, 3 acres permanent meadows, 11 acres woodland forests, 1 acre unimproved land, valued at $6500.
- Farm machinery valued at $1000; livestock, $600; fences, $25; fertilizers, $30; wages paid for farm labor, $30; weeks of hired labor, 6; estimated value of all farm productions, $1000.
- Acres of grass mown, 18; not mown, 8; 30 tons of hay harvested.
- 3 horses, 6 milch cows, 5 other cattle, 6 calves dropped, 4 cattle sold living, 1 cattle slaughtered, 1 cattle died, strayed, stolen and not recovered, 600 lbs. butter, 5 swine, 50 poultry in barnyard, 500 dozen eggs produced.
- 6 acres Indian corn, 250 bu. Indian corn, 8 acres oats, 320 bu. oats, 1/2 acre rye, 12 bu. rye, 8 acres wheat, 210 bu. wheat, 3/4 acres Irish potatoes, 50 bu. Irish potatoes, 1/32 acre sweet potatoes, 3 bu. sweet potatoes.
- 2 acres with 100 apple trees, 10 bu. apples; total value of orchard products, $2.
- 3 cords wood cut valued at $3.

*Federal census lists William, 47; Sarah, 38 (wife); Lizzie, 20; Hattie Ann, 18; Samuel B., 12; Amanda, 9; John S., 1.*

## Joseph B. Good

- Owner
- 70 tilled acres, 13 acres permanent meadows, 32 acres woodland forests, 2 acres unimproved land, valued at $6000.
- Farm machinery valued at $700; livestock, 500; fences, $50; fertilizers, $40; wages paid for farm labor, $50; weeks of hired labor, 8; estimated value of all farm productions, $1200.
- Acres of grass mown, 25; not mown, 10; 40 tons of hay harvested.
- 5 horses, 5 milch cows, 8 other cattle, 7 calves dropped, 8 cattle purchased, 10 cattle sold living, 2 cattle slaughtered, 900 lbs. butter, 5 swine, 40 poultry in barnyard, 500 dozen eggs produced.
- 10 acres Indian corn, 300 bu. Indian corn, 9 acres oats, 390 bu. oats, 2 acres rye, 30 bu. rye, 8 acres wheat, 180 bu. wheat, 1 acre Irish potatoes, 130 bu. Irish potatoes, 1/32 acre sweet potatoes, 4 bu. sweet potatoes, 1 acre tobacco, 900 lbs. tobacco.
- 3 acres with 40 apple trees, 10 bu. apples; 12 peach trees, 2 bu. peaches; total value of orchard products, $5.
- Market gardens valued at $5.
- 10 cords of wood cut valued at $15.

*Federal census lists Joseph, 45; Anna, 41 (wife); Louis G., 19, a teacher; Monroe G., 16; Ezra G., 14; Hattie Ann, 11; Mary G., 2; Nancy, 51 (sister); Jonas, 85 (father).*

## Peter Frankhouser

- Rents for shares of products.
- 50 tilled acres, valued at $2500.
- Farm machinery valued at $500; livestock, $400; fences, $25; fertilizers, 50; wages paid for farm labor, $50; estimated value of all farm productions, $500.
- Acres of grass mown, 8; not mown, 4; 10 tons of hay harvested.
- 3 horses, 4 milch cows, 4 other cattle, 4 calves dropped, 1 cattle slaughtered, 200 lbs. butter, 6 swine, 40 poultry in barnyard, 200 dozen eggs produced.
- 6 acres Indian corn, 150 bu. Indian corn, 6 acres oats, 150 bu. oats, 1/4 acre rye, 5 bu. rye, 6 acres wheat, 100 bu. wheat, 1/2 acre Irish potatoes, 30 bu. Irish potatoes, 2 bu. sweet potatoes, 1/4 acre tobacco, 400 lbs. tobacco.
- 1 acre with 40 apple trees, 5 bu. apples; 6 peach trees, 2 bu. peaches; total value of orchard products, $2.
- 6 cords of wood cut valued at $6.

*Federal census lists Peter, 43; Elizabeth, 43 (wife); Peter, 21; John, 18; Mary A., 13; Ella, 12; Lizzie, 7; Barbara, 6.*

## Henry Miller

- Owner
- 36 tilled acres, 1 acre permanent meadow, valued at $2400.
- Farm machinery valued at $300; livestock, $150; fences, $200; fertilizers, 50; wages paid for farm labor, $50; weeks of hired labor, 8; estimated value of all farm productions, $300.
- Acres of grass mown, 20; not mown, 3; 15 tons of hay harvested.
- 2 horses, 3 milch cows, 2 other cattle, 3 calves dropped, 300 lbs. butter, 4 swine, 28 poultry in barnyard, 150 dozen eggs produced.

- 7 acres Indian corn, 140 bu. Indian corn, 9 acres oats, 180 bu. oats, 1/4 acre rye, 5 bu. rye, 5 acres wheat, 40 bu. wheat, 1/2 acre Irish potatoes, 15 bu. Irish potatoes.
- 1/4 acre with 5 apple trees.

*Federal census lists Henry, 50; Elizabeth, 45 (wife); Jacob V., 14; Anna Mary, 13; Harry V., 12.*

## David H. Sollenberger

- Owner
- 70 tilled acres, 14 acres permanent meadows, 14 acres woodland forests, 5 acres unimproved land, valued at 10,000.
- Farm machinery valued at $1000; livestock, $1000; fences, $100; fertilizers, $120; wages paid for farm labor, $300; weeks of hired labor, 40; estimated value of all farm productions, $1700.
- Acres of grass mown, 35; not mown, 10; 50 tons of hay harvested.
- 3 horses, 6 milch cows, 1 other cattle, 6 calves dropped, 12 cattle purchased, 12 cattle sold living, 1 cattle slaughtered, 60 gal. milk sold, 600 lbs. butter, 5 swine, 30 poultry in barnyard, 200 dozen eggs produced.
- 16 acres Indian corn, 600 bu. Indian corn, 13 acres oats, 500 bu. oats, 4 acres rye, 60 bu. rye, 25 acres wheat, 400 bu. wheat, 3 acres Irish potatoes, 200 bu. Irish potatoes, 2 acres tobacco, 3400 lbs. tobacco.
- 3 acres with 100 apple trees, 20 bu. apples; 1/16 acre with 2 peach trees; total value of orchard products, $5.
- 100 cords of wood cut valued at $150.

*Federal census lists David, 42; Henrietta, 38 (wife); Abraham, 20; Ida, 11; John, 8; Henry, 5; David, 2/12 (Apr.); Martin Coldren, 18 (stepson); Lizzie Coldren, 14 (stepdaughter); Henry Lesley, 27 (brother-in-law); Catherine Lesley, 30 (sister-in-law).*

## J. E. Dougherty

- Rents for shares of products.
- 60 tilled acres; 5 acres permanent meadows, 5 acres unimproved land, valued at $5000.
- Farm machinery valued at $300; livestock, $600; fences, $100; fertilizers, $40; wages paid for farm labor, $40; weeks of hired labor, 6; estimated value of all farm productions, $600.
- Acres of grass mown, 14; not mown, 5; 20 tons of hay harvested.
- 2 horses, 4 milch cows, 4 calves dropped, 300 lbs. butter, 3 swine, 35 poultry in barnyard, 200 dozen eggs produced.
- 9 acres Indian corn, 300 bu. Indian corn, 8 acres oats, 250 bu. oats, 9 acres wheat, 140 bu. wheat, 1/4 acre Irish potatoes, 20 bu. Irish potatoes.

*Federal census lists J. E., 36; Catherine, 27 (wife); Ida, 8; Francis A., 6; Wm. Henry, 4; John A., 2.*

## John Good

- Rents for fixed money rental.
- 18 tilled acres, 6 acres permanent meadows, valued at $1000.
- Farm machinery valued at $200; livestock, $300; fences, $25; wages paid for farm labor, $8; weeks of hired labor, 1; estimated value of all farm productions, $250.
- Acres of grass mown, 6; not mown, 6; 5 tons of hay harvested.
- 1 horse, 4 milch cows, 1 calf dropped, 300 lbs. butter, 50 poultry in barnyard, 250 dozen eggs produced.
- 1/4 acre Indian corn, 25 bu. Indian corn, 1/2 acre oats, 25 bu. oats, 1/2 acre Irish potatoes, 25 bu. Irish potatoes, 1 acre tobacco, 1000 lbs. tobacco.
- 1 acre with 40 apple trees, 20 bu. apples; total value of orchard products, $10.

*Federal census lists John, 46; Sarah, 38 (wife); Mary Ann, 19 (domestic); Abraham, 17; Samuel, 15; Anna, 12; John, 9; Lizzie, 6; Milton, 1; Sarah, 3/12 (Feb.); Jacob Frey, 66, farm laborer; Betsey, 64 (wife).*

## David Musser

- Owner
- 60 tilled acres, 7 acres permanent meadows, 5 acres woodland forests, 7 acres unimproved land, valued at $5000.
- Farm machinery valued at $1200; livestock, $700; fences, $100; fertilizers, $130; wages paid for farm labor, $100; weeks of hired labor, 8; estimated value of all farm productions, $1000.
- Acres of grass mown, 30; not mown, 12; 30 tons of hay harvested, 10 bu. clover seed, 5 bu. grass seed.
- 1 horse, 2 mules and asses, 6 milch cows, 9 other cattle, 6 calves dropped, 4 cattle purchased, 6 cattle sold living, 1 cattle slaughtered, 400 lbs. butter, 5 swine, 50 poultry in barnyard, 3 other poultry, 300 dozen eggs produced.
- 12 acres Indian corn, 300 bu. Indian corn, 12 acres oats, 300 bu. oats, 1 acre rye, 10 bu. rye, 10 acres wheat, 250 bu. wheat, 1 bu. beans (dry), 1 acre Irish potatoes, 135 bu. Irish potatoes, 1/4 acre tobacco, 200 lbs. tobacco.
- 2 acres with 50 apple trees, 10 bu. apples; 10 peach trees, 2 bu. peaches, total value of orchard products, $5.
- 10 cords of wood cut valued at $10.

*Federal census lists David, 51; Hannah, 43 (wife); Sarah, 21; Henry, 18; Martin, 16; Elizabeth, 6; Mary Ann, 1.*

### Samuel Coldren

- Rents for fixed money rental.
- 70 tilled acres, 1 acre permanent meadows, 2 acres woodland forests, 2 acres unimproved land valued at $5000.
- Farm machinery valued at $500; livestock, $600; fences, $100; fertilizers, $50; wages paid for farm labor, $50; weeks of hired labor, 8; estimated value of all farm productions, $500.
- Acres of grass mown, 25; not mown, 10; 25 tons of hay harvested, 2 bu. clover seed, 1 bu. grass seed.
- 3 horses, 6 milch cows, 6 other cattle, 6 calves dropped, 1 cattle purchased, 1 cattle sold living, 1 cattle slaughtered, 500 lbs. butter, 6 swine, 80 poultry in barnyard, 1 other poultry, 300 dozen eggs produced.
- 12 acres Indian corn, 300 bu. Indian corn, 12 acres oats, 300 bu. oats, 1 acre rye, 10 bu. rye, 10 acres wheat, 250 bu. wheat, 1 bu. beans (dry), 1/2 acre Irish potatoes, 25 bu. Irish potatoes.
- 1 acre with 20 apple trees.
- 10 cords of wood cut valued at $10.

*Federal census lists Samuel, 33; Margaret, 34 (wife); Adam, 8; Owen, 7; Barbara, 72 (mother); Solomon, 28 (brother).*

### Jonathan Good

- Owner
- 40 tilled acres, 2 acres permanent meadows, 6 acres woodland forests, 1 acre unimproved land, valued at $5000.
- Farm machinery valued at $400; livestock, $500; fences, $50; fertilizers, $25; wages paid for farm labor, $50; weeks of hired labor, 8; estimated value of all farm productions, $750.
- Acres of grass mown, 20; not mown, 10; 25 tons of hay harvested, 2 bu. clover seed, 1 bu. grass seed.
- 5 horses, 8 milch cows, 4 other cattle, 6 calves dropped, 1 cattle purchased, 1 cattle slaughtered, 1 cattle died, strayed, stolen and not recovered, 500 lbs. butter, 4 swine, 50 poultry in barnyard, 300 dozen eggs produced.
- 8 acres Indian corn, 250 bu. Indian corn, 8 acres oats, 200 bu. oats, 1/4 acre rye, 5 bu. rye, 10 acres wheat, 200 bu. wheat, 1 bu. beans (dry), 1/4 acre Irish potatoes, 25 bu. Irish potatoes, 1/2 acre tobacco, 200 lbs. tobacco.
- 1 acre with 20 apple trees.
- 10 cords of wood cut valued at $10.

*Federal census lists Jonathan, 45; Hettie, 46 (wife); Amanda, 25; Mary Ann, 23; Moses, 20; Sara, 16.*

### Daniel Bauman

- Rents for shares of products.
- 40 tilled acres, 8 acres permanent meadows, 13 acres woodland forests, 4 acres unimproved land, valued at $4000.
- Farm machinery valued at $700; livestock, $500; fences, $100; fertilizers, $50; wages paid for farm labor, $100; weeks of hired labor, 12; estimated value of all farm productions, $600.
- Acres of grass mown, 14; not mown, 5; 20 tons of hay harvested, 1 bu. clover seed, 1 bu. grass seed.
- 3 horses, 6 milch cows, 2 other cattle, 6 calves dropped, 1 cattle slaughtered, 600 lbs. butter, 6 swine, 25 poultry in barnyard, 100 dozen eggs produced.
- 5 acres Indian corn, 200 bu. Indian corn, 5 acres oats, 200 bu. oats, 1/2 acre rye, 15 bu. rye, 6 acres wheat, 125 bu. wheat, 1 acre Irish potatoes, 100 bu. Irish potatoes.
- 2 acres with 25 apple trees.
- 33 cords of wood cut valued at $33.

*Federal census lists Daniel, 29; Mary Ann, 27 (wife); Israel, 6; Harry, 4; Sally, 1; Sarah 19 (sister).*

### Henry Voneidaman

- Owner
- 28 tilled acres, 9 acres permanent meadows, 6 acres woodland forests, 6 acres unimproved land, valued at $3000.
- Farm machinery valued at $500; livestock, $500; fences, $100; fertilizers, $25; wages paid for farm labor, $100; weeks of hired labor, 12; estimated value of all farm productions, $500.
- Acres of grass mown, 14; not mown, 12; 20 tons of hay harvested.
- 2 horses, 4 milch cows, 3 other cattle, 4 calves dropped, 1 cattle slaughtered, 400 lbs. butter, 6 swine, 100 poultry in barnyard, 730 dozen eggs produced.
- 2 acres oats, 50 bu. oats, 1/4 acre rye, 6 bu. rye, 4 acres wheat, 80 bu. wheat, 1/2 acre Irish potatoes, 100 bu. Irish potatoes.

*Federal census lists Henry Vonneida, 54, a miller; Sarah, 45; Barton, 17; Isabella, 21; Nancy, 13; Adam, Jr., 19; Elizabeth Grill, 58 (sister-in-law); David McColm, 14, servant.*

### Joseph Musser

- Owner
- 50 tilled acres, 10 acres permanent meadows, 5 acres woodland forests, 5 acres unimproved land, valued at $4000.

- Farm machinery valued at $600; livestock, $500; fences, $50; wages paid for farm labor, $100; weeks of hired labor, 12; estimated value of all farm productions, $700.
- Acres of grass mown, 15; not mown, 10; 25 tons of hay harvested, 5 bu. grass seed.
- 3 horses, 6 milch cows, 4 other cattle, 6 calves dropped, 3 cattle sold living, 1 cattle slaughtered, 400 lbs. butter, 4 swine, 25 poultry in barnyard, 100 dozen eggs produced.
- 8 acres Indian corn, 200 bu. Indian corn, 8 acres oats, 100 bu. oats, 1 acre rye, 5 bu. rye, 9 acres wheat, 160 bu. wheat, 1/2 acre Irish potatoes, 25 bu. Irish potatoes, 2 acres tobacco, 1050 lbs. tobacco.
- 1 acre with 10 apple trees.
- 10 cords of wood cut valued at $10.

*Federal census lists Joseph, 37; Fanny, 32 (wife); John, 8; Noah, 7; Adam, 5; Solomon, 4; Lydia, 6/12 (Oct.).*

## John Zinn

- Owner
- 40 tilled acres, 8 acres permanent meadows, 5 acres woodland forests, 30 acres unimproved land, valued at $3000.
- Farm machinery valued at $200; livestock, $300; fences, $60; fertilizers, $30; wages paid for farm labor, $20; weeks of hired labor, 2; estimated value of all farm productions, $600.
- Acres of grass mown, 15; not mown, 10; 25 tons of hay harvested, 5 bu. grass seed.
- 2 horses, 4 milch cows, 3 other cattle, 4 calves dropped, 1 cattle sold living, 1 cattle slaughtered, 200 lbs. butter, 2 sheep on hand, 3 fleeces, 6 lbs. wool, 5 swine, 20 poultry in barnyard, 100 dozen eggs produced.
- 5 acres Indian corn; 200 bu. Indian corn, 5 acres oats, 200 bu. oats, 1 acre rye, 5 bu. rye, 5 acres wheat, 100 bu. wheat, 1/2 acre Irish potatoes, 40 bu. Irish potatoes.
- 1/2 acre with 20 apple trees, 20 bu. apples; total value of orchard products, $10.
- 10 cords of wood cut valued at $10.

*Federal census lists John, 56; Lydia, 48 (wife); Jacob, 19; Katie, 12; Pierce John, 8.*

## Samuel G. Seifrit

- Owner
- 23 tilled acres, valued at $1000.
- Farm machinery valued at $300; fences, $20; fertilizers, $30; wages paid for farm labor, $25; weeks of hired labor, 4; estimated value of all farm productions, $200.
- Acres of grass mown, 4; 8 tons of hay harvested.
- 3 acres Indian corn, 50 bu. Indian corn, 5 acres oats, 200 bu. oats, 2 acres wheat, 40 bu. wheat.
- 30 cords of wood cut valued at $45.

*Federal census lists Samuel, 35, a justice of the peace; Mariah, 29 (wife); Edwin Louis, 7; Lizzie, 5; Harry Walter, 8.*

## Isaac Burkholder

- Owner
- 18 tilled acres, 1 acre permanent meadow, 5 acres woodland forests, 3 acres unimproved land, valued at $1800.
- Farm machinery valued at $300; livestock, $800; fences, $100; fertilizers, $25; wages paid for farm labor, $15; weeks of hired labor, 2; estimated value of all farm productions, $400.
- Acres of grass mown, 7; not mown, 3; 8 tons of hay harvested; 1 bu. grass seed.
- 1 horse, 4 milch cows, 2 other cattle, 4 calves dropped, 1 cattle slaughtered, 250 lbs. butter, 4 swine, 25 poultry in barnyard, 300 dozen eggs produced.
- 2 acres Indian corn, 60 bu. Indian corn, 3 acres oats, 65 bu. oats, 1 acre rye, 10 bu. rye, 2 acres wheat, 30 bu. wheat, 1/4 acre Irish potatoes, 30 bu. Irish potatoes.
- 1 acre with 20 apple trees.
- 4 lbs. honey, 4 lbs. beeswax.
- 10 cords of wood cut valued at $10.

*Federal census lists Isaac, 56; Catherine, 62 (wife); Augustus Harting, 16, a servant.*

## Peter B. Good

- Owner
- 45 tilled acres, 222 acres permanent meadows, 65 acres woodland forests, valued at $9000.
- Farm machinery valued at $600; livestock, $1500; fences, $100; fertilizers, $100; wages paid for farm labor, $100; weeks of hired labor, 25; estimated value of all farm productions, $800.
- Acres of grass mown, 18; not mown, 250; 20 tons of hay harvested, 2 bu. clover seed, 11 bu. grass seed.
- 2 horses, 2 milch cows, 2 calves dropped, 16 cattle purchased, 10 cattle sold living, 1 cattle, died, strayed, stolen, and not recovered, 100 lbs. butter, 2 swine, 75 poultry in barnyard, 250 dozen eggs produced.
- 6 acres Indian corn, 200 bu. Indian corn, 7 acres oats, 300 bu. oats, 6 acres wheat, 150 bu. wheat, 1/4 acre Irish potatoes, 25 bu. Irish potatoes, 2 bu. sweet potatoes.
- 2 acres with 40 apple trees, 50 bu. apples; total value of orchard products, $15.

*Federal census lists Peter, 56; Hannah, 53 (wife).*

### Peter B. Good

- Owner
- 89 tilled acres, 8 acres permanent meadows, 5 acres woodland forests, 8 acres unimproved land, valued at $3400.
- Farm machinery valued at $500; livestock, $500; fences, $50; fertilizers, $100; wages paid for farm labor, $150; weeks of hired labor, 30; estimated value of all farm productions, $1200.
- Acres of grass mown, 20, 20 tons of hay harvested, 20 bu. grass seed.
- 4 horses, 2 mules and asses, 4 milch cows, 1 other cattle, 4 calves dropped, 200 lbs. butter, 6 swine, 75 poultry in barnyard, 250 dozen eggs produced.
- 14 acres Indian corn, 600 bu. Indian corn, 10 acres oats, 200 bu. oats, 14 acres wheat, 250 bu. wheat, 1/2 acre Irish potatoes, 50 bu. Irish potatoes, 1 acre tobacco, 1000 lbs. tobacco.
- 4 acres with 40 apple trees.

*This is a second farm.*

### John Werner

- Rents for shares of products.
- 18 tilled acres, 4 acres permanent meadows, valued at $2000.
- Farm machinery valued at $200; livestock, $500; fences, $50; fertilizers, $50; wages paid for farm labor, $25; weeks of hired labor, 4; estimated value of all farm productions, $300.
- Acres of grass mown, 6; not mown, 2, 6 tons of hay harvested.
- 1 horse, 1 milch cow, 1 other cattle, 1 calf dropped, 100 lbs. butter, 4 swine, 60 poultry in barnyard, 200 dozen eggs produced.
- 3 acres Indian corn, 75 bu. Indian corn, 4 acres wheat, 100 bu. wheat, 1/2 acre Irish potatoes, 50 bu. Irish potatoes, 1 1/2 acre tobacco, 1500 lbs. tobacco.
- 1/4 acre with 20 apple trees.

*Federal census lists John, 30; Susanna, 26 (wife); Milton H., 4; John, 2; Agnes, 7/12 (Nov.); Elizabeth Fidler, 73 (mother-in-law).*

### Richard Fichthorn

- Owner
- 28 tilled acres, 4 acres permanent meadows, 3 acres woodland forests, 2 acres unimproved land, valued at $2500.
- Farm machinery valued at $200; livestock, $200; fences, $50; fertilizers, $50; wages paid for farm labor, $25; weeks of hired labor, 4; estimated value of all farm productions, $300.
- Acres of grass mown, 8; not mown, 5, 10 tons of hay harvested.
- 4 horses, 2 milch cows, 2 other cattle, 2 calves dropped, 200 lbs. butter, 2 swine, 50 poultry in barnyard, 200 dozen eggs produced.
- 4 acres Indian corn, 150 bu. Indian corn, 3 acres oats, 20 bu. oats, 1/2 acre rye, 5 bu. rye, 3 acres wheat, 50 bu. wheat, 1/2 acre Irish potatoes, 40 bu. Irish potatoes, 1 acre tobacco, 1000 lbs. tobacco.
- 1/2 acre with 10 apple trees.

*Federal census lists Richard, 57; Mary, 50 (wife); Cyrus, 20; Catherine, 16; Willie, 14; Richard, 9; David Fike, 23, boarder.*

### George Sprecher

- Owner
- 28 tilled acres, 1/2 acre permanent meadows, valued at $2800.
- Farm machinery valued at $500; livestock, $200; fences, $25, fertilizers, $50; wages paid for farm labor, $25; weeks of hired labor, 4; estimated value of all farm productions, $700.
- Acres of grass mown, 10; 15 tons of hay harvested.
- 2 horses, 2 milch cows, 2 calves dropped, 400 lbs. butter, 6 swine, 100 poultry in barnyard, 500 dozen eggs produced.
- 4 acres Indian corn, 200 bu. Indian corn, 1/2 acre rye, 5 bu. rye, 10 acres wheat, 200 bu. wheat, 5 acres Irish potatoes, 100 bu. Irish potatoes, 1 acre tobacco, 1000 lbs. tobacco.
- 1/2 acre with 43 apple trees.

*Federal census lists George, 35; Louisa Jan, 32 (wife); William C., 13; Ulysses G., 11; Henry Y., 10; Lovey Dora, 8; George C., 4; Walter Allen, 2; Oliver Franklin, 7/12 (Nov.).*

### Isaac Good

- Owner
- 49 tilled acres, 4 acres permanent meadows, 6 acres woodland forests, 5 acres unimproved land, valued at $4000.
- Farm machinery valued at $1000; livestock, $300; fences, $50; fertilizers, $45, wages paid for farm labor, $50; weeks of hired labor, 8; estimated value of all farm productions, $500.
- Acres of grass mown, 18; not mown, 12; 12 tons of hay harvested.
- 2 horses, 5 milch cows, 7 other cattle, 2 calves dropped, 1 cattle purchased, 300 lbs. butter, 4 swine, 35 poultry in barnyard, 100 dozen eggs produced.

- 7 acres Indian corn, 200 bu. Indian corn, 7 acres oats, 250 bu. oats, 12 acres wheat, 165 bu. wheat, 1/2 bu. Irish potatoes, 40 bu. Irish potatoes.
- 1/2 acre with 25 apple trees.
- 5 cords of wood cut valued at $5.

*Federal census lists Isaac, 63; Mary, 65 (sister); Matty, 58 (sister); Elizabeth Pleam, 75 (boarder); Henry Griffy, 11 (servant); Lydia Good, 58 (sister).*

## Jacob Coldren

- Owner
- 21 tilled acres, 2 acres permanent meadows, valued at $2000.
- Farm machinery valued at $500; livestock, $200; fences, $25; fertilizers, $25, wages paid for farm labor, $25; weeks of hired labor, 25; estimated value of all farm productions, $400.
- Acres of grass mown, 5; 10 tons of hay.
- 2 horses, 8 milch cows, 1 other cattle, 3 calves dropped, 1 cattle slaughtered, 200 lbs. butter, 2 swine, 50 poultry in barnyard, 100 dozen eggs produced.
- 5 acres Indian corn, 100 bu. Indian corn, 5 acres oats, 150 bu. oats, 1/2 acre rye, 5 bu. rye, 5 acres wheat, 100 bu. wheat, 1/2 acre Irish potatoes, 40 bu. Irish potatoes.
- 1/2 acre with 25 apple trees.
- 5 cords of wood cut valued at $5.

*Federal census lists Jacob, 57; Lizzie, 41 (wife); Abraham, 8 (nephew); Lydia, 66 (sister); Nancy Hudson, 71 (mother-in-law).*

## George Lorah

- Owner
- 40 tilled acres, 1/2 acre permanent meadows, 3 acres woodland forests, 2 acres unimproved land, valued at $3500.
- Farm machinery valued at $500; livestock, $500; fences, $30; fertilizers, $50; wages paid for farm labor, $100; weeks of hired labor, 20; estimated value of all farm productions, $700.
- Acres of grass mown, 12; not mown, 5; 15 tons of hay harvested.
- 2 horses, 4 milch cows, 5 other cattle, 4 calves dropped, 2 cattle purchased, 1 cattle slaughtererd, 500 lbs. butter, 5 swine, 25 poultry in barnyard, 100 dozen eggs produced.
- 6 acres Indian corn, 300 bu. Indian corn, 6 acres oats, 300 bu. oats, 3 acres rye, 70 bu. rye, 4 acres wheat, 96 bu. wheat, 1 acre Irish potatoes, 200 bu. Irish potatoes.
- 1/2 acre with 6 apple trees.

*Federal census lists George, 33; Sarah R., 36 (wife); Mary Geist, 10, servant; John Lorah, 9 (nephew).*

## John Weinhold

- Owner
- 15 tilled acres, 2 acres permanent meadows, 1 acre woodland forests, 1 acre unimproved land, valued at $1500.
- Farm machinery valued at $200; livestock, $100; fences, $20; fertilizers, $20; wages paid for farm labor, $10; weeks of hired labor, 2; estimated value of all farm productions, $200..
- Acres of grass mown, 2; 10 tons of hay harvested.
- 1 horse, 3 milch cows, 1 other cattle, 3 calves dropped, 200 lbs. butter, 3 swine, 25 poultry in barnyard, 100 dozen eggs produced.
- 2 acres Indian corn, 50 bu. Indian corn, 2 acres oats, 100 bu. oats, 4 acres wheat, 100 bu. wheat, 1/2 acre Irish potatoes, 50 bu. Irish potatoes, 1/8 acre tobacco, 200 lbs. tobacco.
- 1/4 acre with 10 apple trees.

*Federal census lists John, 40; Caroline, 39 (wife); Levi, 16; Catherine, 13; Sarah S., 11; Mary Alice, 9; Emerson, 6; Lizzie R., 4; Agnes R., 5/12 (Dec.).*

## William Doutrich

- Rents for shares of products.
- 30 tilled acres, 3 acres permanent meadows, 2 acres unimproved land, valued at $1800.
- Farm machinery valued at $300; livestock, $300; fences, $100; fertilizers, $30; wages paid for farm labor, $50; weeks of hired labor, 8; estimated value of all farm productions, $300.
- Acres of grass mown, 6; not mown, 4; 10 tons of hay harvested.
- 1 horse, 2 milch cows, 3 other cattle, 1 calf dropped, 300 lbs. butter, 4 swine, 15 poultry in barnyard, 100 dozen eggs produced.
- 6 acres Indian corn, 170 bu. Indian corn, 4 acres oats, 125 bu. oats, 1 acre rye, 15 bu. rye, 2 acres wheat, 4 bu. wheat, 1/2 acre Irish potatoes, 50 bu. Irish potatoes.
- 1/2 acre with 20 apple trees, 10 bu. apples; total value of orchard products, $5.

*Federal census lists William, 33; Catherine, 25 (wife); Sally, 9; Catherine, 7; Ella, 3; Charles, 1.*

### Isaac Messner

- Owner
- 60 tilled acres, 5 acres permanent meadows, 1 acre woodland forests, 5 acres unimproved land, valued at $3000.
- Farm machinery valued at $400; livestock, $500; fences, $50; fertilizers, $30; wages paid for farm labor, $25; weeks of hired labor, 4; estimated value of all farm productions, $600.
- Acres of grass mown, 20; not mown, 10; 20 tons of hay harvested, 1 1/2 bu. clover seed, 1 bu. grass seed.
- 4 horses, 5 milch cows, 4 other cattle, 4 calves dropped, 2 cattle purchased, 500 lbs. butter, 2 swine, 50 poultry in barnyard, 300 dozen eggs produced.
- 8 acres Indian corn, 100 bu. Indian corn, 8 acres oats, 200 bu. oats, 2 acres rye, 15 bu. rye, 10 acres wheat, 150 bu. wheat, 1/2 acre Irish potatoes, 25 bu. Irish potatoes, 3/4 acre tobacco, 700 lbs. tobacco.
- 1/4 acre with 10 apple trees.

*Federal census lists Isaac, 33; Catherine, 26 (wife); Roney A., 10; Isaac F., 8.*

### William Ludwig

- Owner
- 20 tilled acres, valued at $1000.
- Farm machinery valued at $100; livestock, $200; fences, $25; fertilizers, $20; wages paid for farm labor, $20; weeks of hired labor, 4; estimated value of all farm productions, $200.
- Acres of grass mown, 5; not mown, 2; 10 tons of hay harvested.
- 3 horses, 1 milch cow, 1 other cattle, 1 calf dropped, 100 lbs. butter, 2 swine, 40 poultry in barnyard, 150 dozen eggs produced.
- 2 acres Indian corn, 50 bu. Indian corn, 2 acres oats, 100 bu. oats, 1 acre rye, 5 bu. rye, 2 acres wheat, 50 bu. wheat, 1 acre Irish potatoes, 25 bu. Irish potatoes, 3/4 acre tobacco, 700 lbs. tobacco.
- 1/4 acre with 10 apple trees.

*Federal census lists only a Philip, 48; Esther, 45 (wife); Jess, 23, farm laborer; John, 22, farm laborer; Prasilla, 21; Adam, 19, farm laborer; William, 18, farm laborer; Henry, 17, farm laborer; Samuel, 16, farm laborer; Susan, 14; Mary, 11; Emeline, 9; Caroline, 6. (1870 census lists William and everyone down to Mary with comparable ages.)*

### Peter Leininger

- Owner
- 8 tilled acres, valued at $1000.
- Farm machinery valued at $100; livestock, $100; fences, $25; fertilizers, 10; wages paid for farm labor, $10; weeks of hired labor, 2; estimated value of all farm productions, $200.
- Acres of grass mown, 2; not mown, 1; 4 tons of hay harvested.
- 2 milch cows, 2 calves dropped, 200 lbs. butter, 3 swine, 50 poultry in barnyard, 200 dozen eggs produced.
- 2 acres Indian corn, 50 bu. Indian corn, 2 acres oats, 100 bu. oats, 1 acre rye, 10 bu. rye, 2 acres wheat, 40 bu. wheat, 1/2 acre Irish potatoes, 25 bu. Irish potatoes.

*Federal census lists Peter, 63, a shoemaker; Catherine, 60 (wife).*

### Allen Brenneman

- Owner
- 18 tilled acres, 1 acre woodland forests, 2 acres unimproved land, valued at $1500.
- Farm machinery valued at $500; livestock, $200; fences, $50; fertilizers, $60; wages paid for farm labor, $25; weeks of hired labor, 6; estimated value of all farm productions, $300.
- Acres of grass mown, 4; not mown, 2; 8 tons of hay harvested, 2 bu. clover seed.
- 1 horse, 1 milch cow, 3 other cattle, 1 calf dropped, 150 lbs. butter, 3 swine, 35 poultry in barnyard, 200 dozen eggs produced.
- 3 acres Indian corn, 100 bu. Indian corn, 1/2 acre oats, 50 bu. oats, 1/2 acre rye, 10 bu. rye, 2 acres wheat, 40 bu. wheat, 1/2 acre Irish potatoes, 40 bu. Irish potatoes.
- 1/2 acre with 18 apple trees.

*Federal census lists Allen, 53, a master mason; Catherine, 53 (wife); Louisa, 17; Lydia, 15; Harry Youndt, 2 (grandchild).*

### Peter Good

- Owner
- 40 tilled acres, 3 acres permanent meadows, 10 acres unimproved land, valued at $3000.
- Farm machinery valued at $500; livestock, $500; fences, $50; fertilizers, $70; wages paid for farm labor, $100; weeks of hired labor, 28; estimated value of all farm productions, $400.
- Acres of grass mown, 4; not mown, 2; 10 tons of hay harvested, 2 bu. clover seed, 2 bu. grass seed.
- 2 horses, 4 milch cows, 2 other cattle, 1 calf dropped, 400 lbs. butter, 5 swine, 75 poultry in barnyard, 300 dozen eggs produced.
- 6 acres Indian corn, 150 bu. Indian corn, 6 acres oats, 265 bu. oats, 1 acre rye, 21 bu. rye, 7 acres wheat, 126 bu. wheat, 1/2 acre Irish potatoes, 40 bu. Irish potatoes.
- 1/2 acre with 22 apple trees.

*Federal census lists Peter, 38; Mary, 29 (wife); Harry, 3; John, 2; Emma, 1.*

## Jacob Steffe

- Owner
- 60 tilled acres, 4 acres permanent meadows, 1 acre woodland forests, 1 acre unimproved land, valued at $4000.
- Farm machinery valued at $600; livestock, $800; fences, $100; fertilizers, $100; wages paid for farm labor, $100; weeks of hired labor, 25; estimated value of all farm productions, $800.
- Acres of grass mown, 10; not mown, 2; 20 tons of hay harvested, 2 bu. clover seed, 2 bu. grass seed.
- 3 horses, 5 milch cows, 7 other cattle, 5 calves dropped, 1 cattle slaughtered, 500 lbs. butter, 5 swine, 50 poultry in barnyard, 400 dozen eggs produced.
- 10 acres Indian corn, 300 bu. Indian corn, 10 acres oats, 300 bu. oats, 2 acres rye, 20 bu. rye, 10 acres wheat, 200 bu. wheat, 1 acre Irish potatoes, 40 bu. Irish potatoes, 1 acre tobacco, 1000 lbs. tobacco.
- 1/2 acre with 25 apple trees, 25 bu. apples; total value of orchard products, $20.
- 10 cords of wood cut valued at $10.

*Federal census lists Jacob, 54; Susanna, 42 (wife); Barbara Ann, 24; Samuel, 21; Cyrus, 17; Emmanial Theo., 15; John Barton, 11; William C., 9; Lydia Ann, 12; Susanna Louisa, 7; Sally Catherine, 4; Caroline, 2; Isaac M. Bouder, 2 (grandchild).*

## Joseph B. Wenger

- Owner
- 21 tilled acres, 4 acres permanent meadows, 2 acres unimproved land, valued at $2000.
- Farm machinery valued at $400; livestock, $200; fences, $50; fertilizers, $50; wages paid for farm labor, $25; weeks of hired labor, 4; estimated value of all farm productions, $400.
- Acres of grass mown, 7; not mown, 4; 7 tons of hay harvested.
- 1 horse, 4 milch cows, 1 calf dropped, 400 lbs. butter, 4 swine, 75 poultry in barnyard, 400 dozen eggs produced.
- 3 acres Indian corn, 100 bu. Indian corn, 2 1/2 acres oats, 57 bu. oats, 6 acres wheat, 84 bu. wheat, 1 acre Irish potatoes, 50 bu. Irish potatoes.
- 1 acre with 25 apple trees, 25 bu. apples; total value of orchard products, $10.

*Federal census does not list J. B. Wenger and family. Due to the manner in which the censuses were taken, the location on the federal census lists the following family as being on this farm: Jacob Hartman, 35, a miller; Catherine, 31 (wife); Anna, 12; Lizzie, 10; Jacob, 69 (father); Caziah, 71 (mother).*

## William Brossman

- Owner
- 60 tilled acres, 2 acres permanent meadows, 15 acres woodland forests, 10 acres unimproved land, valued at $5000.
- Farm machinery valued at $500; livestock, $700; fences, $75; fertilizers, $100; wages paid for farm labor, $100; weeks of hired labor, 20; estimated value of all farm productions, $1000.
- Acres of grass mown, 20; not mown, 8; 20 tons of hay harvested, 2 bu. clover seed, 25 bu. grass seed.
- 5 horses, 7 milch cows, 8 other cattle, 7 calves dropped, 2 cattle slaughtered, 700 lbs. butter, 6 swine, 35 poultry in barnyard, 250 dozen eggs produced.
- 12 acres Indian corn, 300 bu. Indian corn, 10 acres oats, 300 bu. oats, 3 acres rye, 10 bu. rye, 12 acres wheat, 200 bu. wheat, 1/2 acre Irish potatoes, 20 bu. Irish potatoes, 2 acres tobacco, 1300 lbs. tobacco.
- 1/2 acre with 10 apple trees.
- 5 cords of wood cut valued at $5.

*Federal census lists William, 48; Sarah, 44 (wife); John, 22; Hanna E., 18; Susanna, 17; William, 12; Daniel, 9; Henry Snyder, 75, boarder.*

## John W. Overholtzer

- Owner
- 90 tilled acres, 2 acres permanent meadows, 3 acres woodland forests, 5 acres unimproved land, valued at $6000.
- Farm machinery valued at $1500; livestock, $1000, fences, $200, fertilizers, $50, wages paid for farm labor, $150; weeks of hired labor, 25; estimated value of all farm productions, $1500.
- Acres of grass mown, 20; not mown, 12; 30 tons of hay harvested, 4 bu. clover seed, 4 bu. grass seed.
- 8 horses, 4 milch cows, 3 other cattle, 4 calves dropped, 1 cattle purchased, 1 cattle sold living, 1 cattle slaughtered, 500 lbs. butter, 4 sheep on hand, 6 lambs dropped, 4 fleeces, 24 lbs. of wool, 6 swine, 75 poultry in barnyard, 250 dozen eggs produced.
- 12 acres Indian corn, 600 bu. Indian corn, 12 acres oats, 600 bu. oats, 11 acres rye, 110 bu. rye, 20 acres wheat, 400 bu. wheat, 1 acre Irish potatoes, 100 bu. Irish potatoes, 1 acre tobacco, 800 lbs. tobacco.
- 1 acre apple trees.
- 15 cords of wood cut valued at $15.

*Federal census lists John, 61; Lydia, 53 (wife); Anna E., 21; Richard D., 19; Abraham, 17; Barbara E., 15; Samuel David, 67 (brother-in-law; insane).*

## Fianna Kern

- Owner
- 23 tilled acres, valued at $2000.
- Farm machinery valued at $300; livestock, $300; fences, $50; fertilizers, $50; wages paid for farm labor, $25; weeks of hired labor, 6; estimated value of all farm productions, $400.
- Acres of grass mown, 4; not mown, 1; 8 tons of hay harvested.
- 2 horses, 4 milch cows, 1 cattle purchased, 400 lbs. butter, 2 swine, 50 poultry in barnyard, 100 dozen eggs produced.
- 2 acres Indian corn, 50 bu. Indian corn, 2 acres oats, 50 bu. oats, 1 acre rye, 10 bu. rye, 2 acres wheat, 50 bu. wheat, 1 acre Irish potatoes, 50 bu. Irish potatoes, 1 acre tobacco, 1100 lbs. tobacco.
- 1/2 acre with 20 apple trees, 10 bu. apples; total value of orchard products, $5.

*Federal census lists Fianna, widow of Jacob Kern; Edward, 38 (son and farmer); Henry, 33 (son); Richard, 32 (son).*

## Elias Beck

- Owner
- 16 tilled acres, 1 acre permanent meadows, valued at $1000.
- Farm machinery valued at $50; livestock, $200; fences, $50; fertilizers, $25; wages paid for farm labor, $25; weeks of hired labor, 4; estimated value of all farm productions, $500.
- Acres of grass mown, 2; not mown, 2; 5 tons of hay harvested.
- 1 horse, 2 milch cows, 2 calves dropped, 1 cattle sold living, 1 cattle slaughtered, 150 lbs. butter, 1 sheep purchased, 3 swine, 30 poultry in barnyard, 3 other poultry, 150 dozen eggs produced.
- 3 acres Indian corn, 100 bu. Indian corn, 2 acres oats, 50 bu. oats, 3 acres wheat, 50 bu. wheat, 1/2 acre Irish potatoes, 25 bu. Irish potatoes.
- 1/2 acre with 4 apple trees.

*Federal census lists Elias, 55; Sarah, 54 (wife); Susanna, 13; William, 11.*

## Lazarus Moyer

- Owner
- 19 tilled acres, valued at $2000.
- Farm machinery valued at $400; livestock, $ 200; fences, 50; fertilizers, 25; wages paid for farm labor, $25; weeks of hired labor, 4; estimated value of all farm productions, $300.
- Acres of grass mown, 6; 10 tons of hay harvested.
- 1 horse, 3 milch cows, 3 calves dropped, 1 cattle sold living, 1 cattle slaughtered, 300 lbs. butter, 3 swine, 30 poultry in barnyard, 100 dozen eggs produced.
- 2 acres Indian corn, 50 bu. Indian corn, 3 acres oats, 58 bu. oats, 1/2 acre rye, 12 bu. rye, 3 acres wheat, 75 bu. wheat, 1/2 acre Irish potatoes, 70 bu. Irish potatoes, 1/2 acre tobacco, 600 lbs. tobacco.
- 1 acre with 20 apple trees.

*Federal census lists Lazarus, 40; Sarah, 39 (wife); Milton, 15; Elizabeth, 13; Fianna, 11, Ellen, 9; Jacob Fankhouser, 68 (father-in-law), a carpenter.*

## John H. Frey

- Rents for shares of products.
- 35 tilled acres, 7 acres permanent, valued at $2000.
- Farm machinery valued at $400; livestock, $300; fences, $50; fertilizers, $25; wages paid for farm labor, $25; weeks of hired labor, 4; estimated value of all farm productions, $300.
- Acres of grass mown, 6; not mown, 2; 12 tons of hay harvested
- 2 horses, 3 milch cows, 1 other cattle, 1 calf dropped, 300 lbs. butter, 3 swine, 30 poultry in barnyard, 100 dozen eggs produced.
- 5 acres Indian corn, 100 bu. Indian corn, 4 acres oats, 100 bu. oats, 1/2 acre rye, 6 bu. rye, 10 acres wheat, 200 bu. wheat, 1/2 acre Irish potatoes, 25 bu. Irish potatoes.
- 1 acre with 20 apple trees.

*Federal census lists John, 39; Catherine, 31.*

## Jacob Burkhart

- Owner
- 60 tilled acres, 4 acres permanent meadows, 3 acres woodland forests, 11 acres unimproved land, valued at $4000.
- Farm machinery valued at $300; livestock, $500; fences, $50; fertilizers, $50; wages paid for farm labor, $75; weeks of hired labor, 12; estimated value of all farm productions, $500.
- Acres of grass mown, 16; not mown, 11; 18 tons of hay harvested, 6 bu. clover seed.
- 4 horses, 6 milch cows, 6 other cattle, 6 calves dropped, 1 cattle sold living, 1 cattle slaughtered, 400 lbs. butter, 7 swine, 40 poultry in barnyard, 400 dozen eggs produced.
- 9 acres Indian corn, 400 bu. Indian corn, 7 acres oats, 250 bu. oats, 3 acres rye, 20 bu. rye, 14 acres wheat, 275 bu. wheat, 1 acre Irish potatoes, 40 bu. Irish potatoes.

- 2 acres with 20 apple trees, 13 bu. apples; total value of orchard products, $5.
- 12 cords of wood cut valued at $12.

*Federal census lists Jacob, 53; Rebecca, 50 (wife); Louis, 25; Sarah, 22; Henry, 17; Lydia Ellen, 11.*

### Michael Yetter

- Owner
- 9 tilled acres, 3 acres unimproved land, valued at $1200.
- Farm machinery valued at $300; livestock, $300; fences, $50; fertilizers, $50; wages paid for farm labor, $75; weeks of hired labor, 12; estimated value of all farm productions, $500.
- Acres of grass mown, 4; 4 tons of hay harvested.
- 1 horse, 2 milch cows, 1 calf dropped, 1 cattle purchased, 1 cattle sold living, 320 lbs. butter, 9 swine, 40 poultry in barnyard, 350 dozen eggs produced.
- 4 acres wheat, 70 bu. wheat, 1 acre Irish potatoes, 40 bu. Irish potatoes, 1/2 acre tobacco, 412 lbs. tobacco.
- 1/2 acre with 12 apple trees, 5 bu. apples; total value of orchard products, $3.

*Federal census lists Michael, 48, a miller; Louisa, 46 (wife); Mary A., 26, a teacher; Frank J., 24, a teacher; Louisa S., 22, a teacher; Albert E., 10; Charles M., 7.*

### Henry Slabach

- Owner
- 23 tilled acres, valued at $1600.
- Farm machinery valued at $300; livestock, $200; fences, $20; fertilizers, $60; estimated value of all farm productions, $100.
- Acres of grass mown, 8; 8 tons of hay harvested.
- 3 horses, 3 milch cows, 4 other cattle, 3 calves dropped, 1 cattle purchased, 1 cattle sold living, 1 cattle slaughtered, 300 lbs. butter, 3 swine, 50 poultry in barnyard, 1 other poultry, 360 dozen eggs produced.
- 4 acres Indian corn, 200 bu. Indian corn, 1 acre oats, 25 bu. oats, 5 acres wheat, 90 bu. wheat, 1/2 acre Irish potatoes, 50 bu. Irish potatoes, 3/4 acre tobacco, 600 lbs. tobacco.
- 1/2 acre with 25 apple trees, 25 bu. apples; total value of orchard products, $12.

*Federal census lists Henry, 51; Anna, 42 (wife); Isaac, 15.*

### George Dietrich

- Owner
- 29 tilled acres, 3 acres permanent meadows, 6 acres unimproved land, valued at $1500.
- Farm machinery valued at $300; livestock, $300; fences, $60; fertilizers, $60; wages paid for farm labor, $60; weeks of hired labor, 10; estimated value of all farm productions, $500.
- Acres of grass mown, 10; not mown, 4; 10 tons of hay harvested.
- 2 horses, 3 milch cows, 3 other cattle, 3 calves dropped, 1 cattle slaughtered, 200 lbs. butter, 5 swine, 40 poultry in barnyard, 200 dozen eggs produced.
- 5 acres Indian corn, 80 bu. Indian corn, 5 acres oats, 125 bu. oats, 3/4 acre rye, 5 bu. rye, 9 acres wheat, 100 bu. wheat, 1/4 acre Irish potatoes, 10 bu. Irish potatoes, 1/2 acre tobacco, 800 lbs. tobacco.
- 1 acre with 20 apple trees; 20 bu. apples; total value of orchard products, $4.

*Federal census lists George, 42; Sophia, 33 (wife); Mary Ann, 14; Alice, 9; Addison, 10/12 (Aug.).*

### Martin H. Joh

- Owner
- 8 tilled acres, 2 acres permanent meadows, 6 acres unimproved land, valued at $1000.
- Farm machinery valued at $200; livestock, $200; fences, $50; fertilizers, $30; wages paid for farm labor, $10; weeks of hired labor, 2; estimated value of all farm productions, $75.
- Acres of grass mown, 3; not mown, 2; 1 ton of hay harvested.
- 2 horses, 1 milch cow, 1 calf dropped, 50 lbs. butter, 4 swine, 40 poultry in barnyard, 200 dozen eggs produced.
- 3 acres Indian corn, 70 bu. Indian corn, 3 acres rye, 12 bu. rye, 1/4 acre Irish potatoes, 10 bu. Irish potatoes.

*Federal census lists Martin, 38; Catherine, 34 (wife); Charles, 4; Ada V., 3.*

### Israel Messner

- Owner
- 14 tilled acres, 1 acre permanent meadows, 1 acre unimproved land, valued at $800.
- Farm machinery valued at $100; livestock, $100; fences, $25; fertilizers, $10; wages paid for farm labor, $10; weeks of hired labor, 2; estimated value of all farm productions, $300.
- Acres of grass mown, 2; not mown, 2; 4 tons of hay harvested.
- 1 horse, 1 milch cow, 3 other cattle, 2 calves dropped, 200 lbs. butter, 2 swine, 30 poultry in barnyard, 200 dozen eggs produced.
- 2 acres Indian corn, 100 bu. Indian corn, 1/4 acre oats, 10 bu. oats, 1/2 acre rye, 10 bu. rye, 1 acre wheat, 34 bu. wheat, 1/4 acre Irish potatoes, 10 bu. Irish potatoes, 1 acre tobacco, 2200 lbs. tobacco.
- 1/2 acre with 25 apple trees; 30 bu. apples; total value of orchard products, $15.

• 4 cords of wood cut valued at $4.
*Federal census lists Israel, 48; Lydia, 47 (wife); Sidney M., 21 (daughter).*

### Edward Leise
• Owner
• 20 tilled acres, 4 acres woodland forests, valued at $800.
• Farm machinery valued at $100; livestock, $300; fences, $50; fertilizers, 25; wages paid for farm labor, $10; weeks of hired labor, 2; estimated value of all farm productions, $300.
• Acres of grass mown, 2; not mown, 2; 5 tons of hay harvested.
• 2 horses, 2 milch cows, 1 other cattle, 2 calves dropped, 1 cattle slaughtered, 250 lbs. butter, 3 swine, 25 poultry in barnyard, 150 dozen eggs produced.
• 4 acres Indian corn, 100 bu. Indian corn, 1 acre oats, 25 bu. oats, 1/4 acre rye, 5 bu. rye, 2 acres wheat, 50 bu. wheat, 1/4 acre Irish potatoes, 20 bu. Irish potatoes, 1/2 acre tobacco, 800 lbs. tobacco.
• 1/4 acre with 10 apple trees, 20 bu. apples; total value of orchard products, $5.
• 4 cords of wood cut valued at $4.
*Federal census lists Edward, 53; Elizabeth, 48 (wife); Oliver, 18; Mary Ann, 13.*

### William Wolf
• Owner
• 20 tilled acres, 1 acre permanent meadows, 2 acres unimproved land, valued at $1000.
• Farm machinery valued at $100; livestock, $200; fences, $100; fertilizers, $50; wages paid for farm labor, $10; weeks of hired labor, 2; estimated value of all farm productions, $400.
• Acres of grass mown, 6; not mown, 6 tons of hay harvested.
• 2 horses, 2 milch cows, 5 other cattle, 2 calves dropped, 1 cattle slaughtered, 250 lbs. butter, 1 sheep purchased, 3 swine, 30 poultry in barnyard, 1 other poultry, 200 dozen eggs produced.
• 3 acres Indian corn, 100 bu. Indian corn, 2 acres oats, 12 bu. oats, 1/4 acre rye, 5 bu. rye, 2 acres wheat, 35 bu. wheat, 1/2 acre Irish potatoes, 25 bu. Irish potatoes, 1/2 acre tobacco, 700 lbs. tobacco.
• 1/2 acre with 25 apple trees, 5 bu. apples; total value of orchard products, $2.
• 2 cords of wood cut valued at $2.
*Federal census lists William, 68; Caziah, 57 (wife); Jonathan, 20; William, 8 (grandson); Catherine Marshall, 79 (mother-in-law).*

### Daniel Weinhold
• Owner
• 20 tilled acres, 2 acres permanent meadows, 8 acres unimproved land, valued at $800.
• Farm machinery valued at $200; livestock, $200; fences, $100; fertilizers, $50; wages paid for farm labor, $10; weeks of hired labor, 2; estimated value of all farm productions, $400.
• Acres of grass mown, 5; 6 tons of hay harvested.
• 2 horses, 3 milch cows, 3 calves dropped, 1 cattle slaughtered, 300 lbs. butter, 2 swine, 75 poultry in barnyard, 400 dozen eggs produced.
• 5 acres Indian corn, 125 bu. Indian corn, 2 acres oats, 50 bu. oats, 4 acres wheat, 100 bu. wheat, 1 acre Irish potatoes, 100 bu. Irish potatoes, 1/2 acre tobacco, 700 lbs. tobacco.
• 1/2 acre with 15 apple trees, 5 bu. apples; total value of orchard products, $2.
• 2 cords of wood cut valued at $2.
*Federal census lists Daniel, 56; Susanna, 49 (wife); George Frey, 79 (Susanna's cousin).*

### Elias Hornberger
• Rents for shares of products.
• 50 tilled acres, 2 acres permanent meadows, 4 acres woodland forests, 8 acres unimproved land, valued at $2000.
• Farm machinery valued at $200; livestock, $400; fences, $100; fertilizers, $50; wages paid for farm labor, $50; weeks of hired labor, 8; estimated value of all farm productions, $300.
• Acres of grass mown, 6; 6 tons of hay harvested.
• 2 horses, 4 milch cows, 1 other cattle, 4 calves dropped, 1 cattle slaughtered, 700 lbs. butter, 3 swine, 50 poultry in barnyard, 200 dozen eggs produced.
• 5 acres Indian corn, 100 bu. Indian corn, 4 acres oats, 100 bu. oats, 1 acre rye, 20 bu. rye, 4 acres wheat, 100 bu. wheat, 1 acre Irish potatoes, 50 bu. Irish potatoes.
• 1 acre with 15 apple trees, 5 bu. apples; total value of orchard products, $2.
• 10 cords of wood cut valued at $10.
*Federal census lists Elias, 30; Adeline, 27 (wife); Samuel, 7; Anna M., 5; David, 3; Amanda, 11/12 (July).*

### Reuben E. Shober
• Owner
• 25 tilled acres, 1 acre permanent meadows, 4 acres permanent woodlands, 30 acres unimproved land, valued at $2750.

- Farm machinery valued at $200; livestock, $200; fences, $50; fertilizers, $40; wages paid for farm labor, $20; weeks of hired labor, 120; estimated value of all farm productions, $600.
- Acres of grass mown, 8; not mown, 10; 16 tons of hay harvested.
- 2 horses, 5 milch cows, 7 other cattle, 4 calves dropped, 1 cattle slaughtered, 600 lbs. butter, 5 swine, 26 poultry in barnyard, 600 dozen eggs produced.
- 3 acres Indian corn, 120 bu. Indian corn, 3 acres oats, 120 bu. oats, 4 acres wheat, 80 bu. wheat, 1/4 acre Irish potatoes, 20 bu. Irish potatoes, 1 acre tobacco, 1400 lbs. tobacco.
- 1 acre with 20 apple trees; 5 bu. apples; total value of orchard products, $2.
- 5 cords of wood cut valued at $5.

*Federal census lists Reuben, 58, a miller; Susanna, 33 (wife); George O., 17; Margaret A., 10; Samuel S., 9; Mary E., 7; John Henry, 5; Elisabeth S., 2; Catherine 4/12 (Feb.); Susanna, 77 (mother).*

### John H. Frey

- Owner
- 50 tilled acres, 2 acres permanent meadows, 1 acre woodland forests, 35 acres unimproved land, valued at $3000.
- Farm machinery valued at $500; livestock, $400; fences, $50; fertilizers, $60; wages paid for farm labor, $100; weeks of hired labor, 20; estimated value of all farm productions, $800.
- Acres of grass mown, 25; not mown, 10; 25 tons of hay harvested.
- 3 horses, 4 milch cows, 6 other cattle, 4 calves dropped, 1 cattle slaughtered, 400 lbs. butter, 4 swine, 35 poultry in barnyard, 260 dozen eggs produced.
- 12 acres Indian corn, 200 bu. Indian corn, 10 acres oats, 400 bu. oats, 15 acres wheat, 200 bu. wheat, 1/2 acre Irish potatoes, 30 bu. Irish potatoes.
- 1 acre with 15 apple trees, 10 bu. apples; total value of orchard products, $2.
- 5 cords of wood cut valued at $5.

*Federal census lists John, 46; Eliza, 37 (wife); Susan, 17; John, 10.*

### David Meiskey

- Rents for fixed money rental.
- 50 tilled acres, 10 acres woodland forests, 20 acres unimproved land, valued at $5000.
- Farm machinery valued at $500; livestock, $500; fences, $50; fertilizers, $50; wages paid for farm labor, $100; weeks of hired labor, 25; estimated value of all farm productions, $800.
- Acres of grass mown, 14; not mown, 6; 20 tons of hay harvested.
- 3 horses, 7 milch cows, 8 other cattle, 7 calves dropped, 600 lbs. butter, 4 swine, 50 poultry in barnyard, 400 dozen eggs produced.
- 9 acres Indian corn, 400 bu. Indian corn, 9 acres oats, 250 bu. oats, 9 acres wheat, 175 bu. wheat, 1/2 acre Irish potatoes, 100 bu. Irish potatoes.
- 3 acres with 30 apple trees; total value of orchard products, $5.
- 10 lbs. honey; 1 lb. beeswax.
- 15 cords of wood cut valued at $15.

*Federal census lists David, 52; Mary, 48 (wife); Margaret Henry, 71 (mother-in-law); Isaac Folk, 14 (hired boy).*

### John Yundt

- Owner
- 15 tilled acres, valued at $1500.
- Farm machinery valued at $200; livestock, $400; fences, $50; fertilizers, $50; wages paid for farm labor, $50; weeks of hired labor, 6; estimated value of all farm productions, $400.
- Acres of grass mown, 5; 10 tons of hay harvested.
- 1 horse, 2 milch cows, 2 other cattle, 2 calves dropped, 1 cattle slaughtered, 250 lbs. butter, 4 swine, 30 poultry in barnyard, 200 dozen eggs produced.
- 2 acres Indian corn, 100 bu. Indian corn, 2 acres oats, 100 bu. oats, 2 acres wheat, 40 bu. wheat, 1/2 acre Irish potatoes, 20 bu. Irish potatoes, 1 acre tobacco, 1,000 lbs. tobacco.
- 2 acres with 20 apple trees.
- 2 cords of wood cut valued at $2.

*Federal census lists John, 65; Sarah, 63 (wife); Lydia A., 27, cigar maker; Catherine, 19, cigar maker; Samuel, 21, butcher; Emma, 21 (daughter-in-law).*

### Israel Hoffman

- Owner
- 80 tilled acres, 3 acres permanent meadows, 11 acres woodland forests, 6 acres unimproved land, valued at $5000.
- Farm machinery valued at $700; livestock, $800; fences, $100; fertilizers, $132; wages paid for farm labor, $150; weeks of hired labor, 40; estimated value of all farm productions, $800.
- Acres of grass mown, 27; not mown, 7; 27 tons of hay harvested.

- 7 horses, 5 milch cows, 7 other cattle, 5 calves dropped, 4 cattle purchased, 4 cattle sold living, 2 cattle slaughtered, 450 lbs. butter, 7 swine, 150 poultry in barnyard, 500 dozen eggs produced.
- 12 acres Indian corn, 400 bu. Indian corn, 10 acres oats, 300 bu. oats, 1 acre rye, 9 bu. rye, 12 acres wheat, 175 bu. wheat, 3/4 acre Irish potatoes, 40 bu. Irish potatoes.
- 1/2 acre with 15 apple trees, 7 peach trees; total value of orchard products, $1.
- 7 cords of wood cut valued at $7.

*Federal census lists Israel, 36; Lydia, 34 (wife); Mary Berkley, 26, servant.*

## Samuel Stover

- Owner
- 50 tilled acres, 3 acres permanent meadows, 1 acre woodland forests, 5 acres unimproved land, valued at $3500.
- Farm machinery valued at $400; livestock, $500; fences, $100; fertilizers, $50; wages paid for farm labor, $50; weeks of hired labor, 12; estimated value of all farm productions, $700.
- Acres of grass mown, 12; not mown, 5; 20 tons of hay harvested.
- 3 horses, 6 milch cows, 5 other cattle, 5 calves dropped, 1 cattle slaughtered, 600 lbs. butter, 5 swine, 40 poultry in barnyard, 300 dozen eggs produced.
- 10 acres Indian corn, 250 bu. Indian corn, 10 acres oats, 200 bu. oats, 1/2 acre rye, 5 bu. rye, 10 acres wheat, 150 bu. wheat, 1/2 acre Irish potatoes, 20 bu. Irish potatoes.
- 1/2 acre with 20 apple trees, 2 peach trees.
- 5 cords of wood cut valued at $5.

*Federal census lists Samuel, 50; Rebecca, 40 (wife); Monroe, 28 (son); Anna Mary, 20; Frances, 18, cigar maker (daughter); Samuel, 3.*

## John Lorah

- Owner
- 40 tilled acres, 3 acres permanent meadows, 4 acres woodland forests, 25 acres unimproved land, valued at $2500.
- Farm machinery valued at $500; livestock, $400; fences, $100; fertilizers, $84; wages paid for farm labor, $30; weeks of hired labor, 4; estimated value of all farm productions, $400.
- Acres of grass mown, 15; not mown, 5; 15 tons of hay harvested.
- 2 horses, 4 milch cows, 3 other cattle, 2 calves dropped, 1 cattle slaughtered, 1 cattle died, strayed, stolen, and not recovered, 400 lbs. butter, 4 swine, 15 poultry in barnyard, 200 dozen eggs produced.
- 5 acres Indian corn, 125 bu. Indian corn, 5 acres oats, 100 bu. oats, 1/4 acre rye, 3 bu. rye, 6 acres wheat, 60 bu. wheat, 3/8 acres Irish potatoes, 55 bu. Irish potatoes.
- 1/2 acre with 15 apple trees, 8 peach trees; total value of orchard products, $5.
- 5 cords of wood cut valued at $5.

*Federal census lists John, 43; Susanna, 48 (wife); John, 20; Anna Louisa, 16.*

## John Wolfkill

- Owner
- 6 tilled acres, valued at $500.
- Farm machinery valued at $100; livestock, $50; fences, $10; fertilizers, $25; estimated value of all farm productions, $200.
- Acres of grass mown, 2; 2 tons of hay harvested.
- 1 milch cow, 150 lbs. butter, 2 swine, 20 poultry in barnyard, 100 dozen eggs produced.
- 1 acre Indian corn, 24 bu. Indian corn, 1/4 acre oats, 7 bu. oats, 1 1/2 acres wheat, 30 bu. wheat, 1/4 acre Irish potatoes, 25 bu. Irish potatoes.

*Federal census lists John, 38; Elizabeth, 36 (wife); Lydia Ann, 13; Catherine, 8.*

## Michael Weinhold

- Owner
- 50 tilled acres, 3 acres permanent meadows, 5 acres woodland forests, 25 acres unimproved land, valued at $4000.
- Farm machinery valued at $500; livestock, $400; fences, $100; fertilizers, $60; wages paid for farm labor, $100; weeks of hired labor, 12; estimated value of all farm productions, $500.
- Acres of grass mown, 12; not mown, 5; 20 tons of hay harvested.
- 3 horses, 1 milch cow, 4 other cattle, 7 calves dropped, 2 cattle sold living, 2 cattle slaughtered, 500 lbs. butter, 5 swine, 50 poultry in barnyard, 400 dozen eggs produced.
- 10 acres Indian corn, 400 bu. Indian corn, 6 acres oats, 150 bu. oats, 9 acres wheat, 175 bu. wheat, 1 acre Irish potatoes, 40 bu. Irish potatoes.
- 1/2 acre with 6 apple trees.
- Market gardens valued at $10.
- 5 cords of wood cut valued at $5.

*Federal census lists Michael, 48; Louisa, 44 (wife); Mary A., 20; Frank, 19; Amanda, 13; Martin, 9.*

### William Voneida

- Owner
- 25 tilled acres, valued at $2500.
- Farm machinery valued at $200; livestock, $200; fences, $50; fertilizers, $25, wages paid for farm labor, $25; weeks of hired labor, 4; estimated value of all farm productions, $200.
- Acres of grass mown, 4; 5 tons of hay harvested.
- 1 horse, 3 milch cows, 3 other cattle, 3 calves dropped, 1 cattle slaughtered, 300 lbs. butter, 3 swine, 40 poultry in barnyard, 300 dozen eggs produced.
- 4 acres Indian corn, 100 bu. Indian corn, 2 acres oats, 50 bu. oats, 3 acres wheat, 50 bu. wheat, 1/2 acre Irish potatoes, 20 bu. Irish potatoes.
- 1 acre with 10 apple trees.
- 2 cords of wood cut valued at $2.

*Federal census lists William, 60; Eliza, 58 (wife); Ezra, 20; Charles Ruth, 6 (grandson).*

### Charles Kraemer

- Owner
- 50 tilled acres, 6 acres permanent meadows, 6 acres woodland forests, 10 acres unimproved land, valued at $2000.
- Farm machinery valued at $800; livestock, $600; fences, $100; fertilizers, $50; wages paid for farm labor, $50; weeks of hired labor, 8; estimated value of all farm productions, $400.
- Acres of grass mown, 25; not mown, 20; 30 tons of hay harvested.
- 2 horses, 5 milch cows, 8 other cattle, 5 calves dropped, 3 cattle sold living, 1 cattle slaughtered, 2 cattle died, strayed, stolen, and not recovered, 500 lbs. butter, 5 swine, 80 poultry in barnyard, 400 dozen eggs produced.
- 10 acres Indian corn, 200 bu. Indian corn, 2 acres oats, 300 bu. oats, 8 acres wheat, 120 bu. wheat, 1/2 acre Irish potatoes, 80 bu. Irish potatoes.
- 6 acres with 60 apple trees.
- 3 cords of wood cut valued at $3.

*Federal census lists Charles, 50; Sophia, 46 (wife); Anna, 21; Sophia, 19; Mary, 18; Theresa, 17; Charles, 14; Amanda, 8.*

### Uriah Messner

- Owner
- 30 tilled acres, 5 acres permanent meadows, 2 acres woodland forests, 28 acres unimproved land, valued at $1500.
- Farm machinery valued at $500; livestock, $500; fences, $50; fertilizers, $50; wages paid for farm labor, $50; weeks of hired labor, 8; estimated value of all farm productions, $200.
- Acres of grass mown, 12; not mown, 20; 20 tons of hay harvested.
- 2 horses, 6 milch cows, 2 other cattle, 6 calves dropped, 1 cattle sold living, 1 cattle slaughtered, 400 lbs. butter, 4 swine, 8 poultry in barnyard, 100 dozen eggs produced.
- 5 acres Indian corn, 150 bu. Indian corn, 5 acres oats, 150 bu. oats, 5 acres wheat, 55 bu. wheat, 1/2 acre Irish potatoes, 25 bu. Irish potatoes.
- 1 acre with 8 apple trees.
- 2 cords of wood cut valued at $2.

*Federal census lists Tobias Messner, 26; Amanda, 23 (wife); Henry, 4; John, 2; Alice, 7/12 (Dec.); Mary Kind, 65 (mother-in-law).*

### Benneville Pennebacker

- Owner
- 20 tilled acres, 1 acre permanent meadows, 2 acres woodland forests, 17 acres unimproved land, valued at $1000.
- Farm machinery valued at $300; livestock, $500; fences, $50; fertilizers, $50; wages paid for farm labor, $25; weeks of hired labor, 6; estimated value of all farm productions, $300.
- Acres of grass mown, 4; not mown, 2; 10 tons of hay harvested.
- 2 horses, 3 milch cows, 2 other cattle, 3 calves dropped, 1 cattle slaughtered, 300 lbs. butter, 3 swine, 25 poultry in barnyard, 100 dozen eggs produced.
- 5 acres Indian corn, 100 bu. Indian corn, 2 acres oats, 50 bu. oats, 5 acres wheat, 50 bu. wheat, 1/2 acre Irish potatoes, 25 bu. Irish potatoes.
- 1 acre with 10 apple trees.

*Federal census lists Benneville, 51; Catherine, 40 (wife).*

### Levi Siverling

- Rents for fixed money rental.
- 80 tilled acres, 6 acres permanent meadows, 10 acres woodland forests, 20 acres unimproved land, valued at $6000.

- Farm machinery valued at $1000; livestock, $500; fences, $100; fertilizers, $100; wages paid for farm labor, $100; weeks of hired labor, 20; estimated value of all farm productions, $1000.
- Acres of grass mown, 30; not mown, 8; 30 tons of hay harvested.
- 3 horses, 6 milch cows, 5 other cattle, 6 calves dropped, 600 lbs. butter, 6 swine, 50 poultry in barnyard, 300 dozen eggs produced.
- 20 acres Indian corn, 500 bu. Indian corn, 10 acres oats, 300 bu. oats, 3 acres rye, 30 bu. rye, 20 acres wheat, 300 bu. wheat, 1/2 acre Irish potatoes, 25 bu. Irish potatoes.
- 1 acre with 30 apple trees, 12 bu. apples; total value of orchard products, $20.
- 12 cords of wood cut valued at $10.

*Federal census lists Levi Seiverling, 34; Mary, 27 (wife); Harry, 9; Martin, 7; Willie, 5; Adam, 2; Matty Druckenbroch, 18 (sister-in-law).*

## William Kline

- Owner
- 55 tilled acres, 1 acre permanent meadows, 1 acre woodland forests, 15 acres unimproved land, valued at $1500.
- Farm machinery valued at $200; livestock, $300; fences, $50; fertilizers, $100; wages paid for farm labor, $130; weeks of hired labor, 20; estimated value of all farm productions, $400.
- Acres of grass mown, 11; not mown, 5; 2 tons of hay harvested.
- 2 horses, 3 milch cows, 2 other cattle, 3 calves dropped, 1 cattle sold living, 300 lbs. butter, 3 swine, 60 poultry in barnyard, 300 dozen eggs produced.
- 16 acres Indian corn, 200 bu. Indian corn, 4 acres oats, 80 bu. oats, 2 acres rye, 2 bu. rye, 7 acres wheat, 25 bu. wheat, 1/2 acre Irish potatoes, 30 bu. Irish potatoes.
- 1 acre with 25 apple trees, 6 bu. apples; total value of orchard products, $20.

*Federal census lists William, 48; Lydia, 36 (wife); Rachel E., 6 (adopted).*

## Samuel Becker

- Owner
- 70 tilled acres, 5 acres permanent meadows, 10 acres woodland forests, 5 acres unimproved land, valued at $6000.
- Farm machinery valued at $1000; livestock, $800; fences, $100; fertilizers, $108; wages paid for farm labor, $40; weeks of hired labor, 6; estimated value of all farm productions, $700.
- Acres of grass mown, $25; not mown, 4; 25 tons of hay harvested.
- 2 horses, 5 milch cows, 8 other cattle, 5 calves dropped, 2 cattle purchased, 1 cattle slaughtered, 1 cattle died, strayed, stolen, and not recovered, 300 lbs. butter, 4 swine, 45 poultry in barnyard, 250 dozen eggs produced.
- 13 acres Indian corn, 350 bu. Indian corn, 9 acres oats, 100 bu. oats, 2 acres rye, 17 bu. rye, 11 acres wheat, 120 bu. wheat, 1/2 acre Irish potatoes, 30 bu. Irish potatoes.
- 2 acres with 12 apple trees, 12 bu. apples; total value of orchard products, $20.
- 10 cords of wood cut valued at $10.

*Federal census lists Samuel, 46; Mary, 45; David, 23; Lydia, 19; Henry, 14; Barbara, 11; Samuel, 9; Benjamin, 6; Mary, 1.*

## Daniel Weinhold

- Owner
- 16 tilled acres, 4 acres permanent meadows, 1 acre unimproved land, valued at $1000.
- Farm machinery valued at $200; livestock, $200; fences, $30; fertilizers, $30; estimated value of all farm productions, $250.
- Acres of grass mown, 5; 5 tons of hay harvested.
- 2 horses, 2 milch cows, 1 other cattle, 2 calves dropped, 1 cattle slaughtered, 200 lbs. butter. 2 swine, 30 poultry in barnyard, 200 dozen eggs produced.
- 3 acres Indian corn, 100 bu. Indian corn, 3 acres oats, 80 bu. oats, 1/4 acre rye, 4 bu. rye, 3 acres wheat, 60 bu. wheat, 1/2 acre Irish potatoes, 25 bu. Irish potatoes.
- 1/16 acre with 12 apple trees, 12 bu. apples; total value of orchard products, $20.

*Federal census lists Daniel, 37; Mary, 31 (wife); Elizabeth Coldren (mother-in-law).*

## Peter Marks

- Owner
- 80 tilled acres, 6 acres permanent meadows, 2 acres woodland forests, 20 acres unimproved land, valued at $7000.
- Farm machinery valued at $1000; livestock, $1000; fences, $100; fertilizers, $100; wages paid for farm labor, $5; weeks of hired labor, 1; estimated value of all farm productions, $1200.
- Acres of grass mown, 30; not mown, 10; 50 tons of hay harvested, 5 bu. clover seed.
- 7 horses, 5 milch cows, 16 other cattle, 5 calves dropped, 6 cattle purchased, 3 cattle sold living, 2 cattle slaughtered, 1 cattle died, strayed, stolen, and not recovered, 400 lbs. butter, 5 swine, 40 poultry in barnyard, 200 dozen eggs produced.

• 12 acres Indian corn, 400 bu. Indian corn, 13 acres oats, 400 bu. oats, 3 acres rye, 40 bu. rye, 11 acres wheat, 250 bu. wheat, 1 acre Irish potatoes, 150 bu. Irish potatoes.
• 4 acres with 40 bu. apple trees, 25 peach trees.

*Federal census lists Peter, 50; Emma, 50 (wife); Edward, 20; Catherine, 19; Joseph, 16; Emaline, 14; John, 13.*

### John Miller
• Owner
• 16 tilled acres, valued at $1600.
• Farm machinery valued at $100; livestock, $200; fences, $25; fertilizers, $25; wages paid for farm labor, $10; weeks of hired labor, 2; estimated value of all farm productions, $500.
• Acres of grass mown, 3; 6 tons of hay harvested.
• 3 milch cows, 3 calves dropped, 300 lbs. butter, 1 sheep purchased, 2 swine, 12 poultry in barnyard, 100 dozen eggs produced.
• 3 acres Indian corn, 100 bu. Indian corn, 3 acres oats, 100 bu. oats, 3 acres wheat, 75 bu. wheat, 1/4 acre Irish potatoes, 25 bu. Irish potatoes.
• 12 apple trees, 2 peach trees.

*Federal census lists John, 46, a carpenter; Catherine, 41 (wife); Ida, 18; Samuel B., 9.*

### Christian Stauffer
• Owner
• 30 tilled acres, valued at $4000.
• Farm machinery valued at $150; livestock, $400; fences, $25; fertilizers, $20; wages paid for farm labor, $50; weeks of hired labor, 8, estimated value of all farm productions, $500.
• Acres of grass mown, 10; 20 tons of hay harvested, 2 bu. clover seed.
• 2 horses, 2 milch cows, 2 calves dropped, 300 lbs. butter, 4 swine, 23 poultry in barnyard, 150 dozen eggs produced.
• 5 acres Indian corn, 200 bu. Indian corn, 5 acres oats, 200 bu. oats, 6 acres wheat, 110 bu. wheat, 1/2 acre Irish potatoes, 40 bu. Irish potatoes.
• 1/4 acre with 12 apple trees.

*Federal census lists Christian, 57, a butcher; Juliann, 66 (wife); Israel L., 25; Ellen, 20 (wife); Alice, 3.*

### Elias Ziegler
• Owner
• 22 tilled acres, valued at $3000.
• Farm machinery valued at $150; livestock, $400; fences, $50; fertilizers, $25; wages paid for farm labor, $25; weeks of hired labor, 8; estimated value of all farm productions, $500.
• Acres of grass mown, 8; 16 tons of hay harvested.
• 2 mules and asses, 3 milch cows, 3 calves dropped, 400 lbs. butter, 6 swine, 38 poultry in barnyard, 200 dozen eggs produced.
• 4 acres Indian corn, 200 bu. Indian corn, 3 acres oats, 100 bu. oats, 5 acres wheat, 100 bu. wheat, 1/4 acre Irish potatoes, 30 bu. Irish potatoes.
• 1/4 acre with 12 apple trees, 2 peach trees.

*Federal census lists Elias, 48; Mary, 42 (wife); Milton, 23; Susan, 18; Anna, 15.*

### Gerson L. Bauman
• Rents for shares of products.
• 53 tilled acres, 10 acres permanent meadows, 20 acres woodland forests, 5 acres unimproved land, valued at $6000.
• Farm machinery valued at 600; livestock, $300; fences, $200; fertilizers, $100; wages paid for farm labor, $125; weeks of hired labor, 20; estimated value of all farm productions, $800.
• Acres of grass mown, 20 tons of hay harvested, 4 bu. clover seed.
• 2 horses, 2 mules and asses, 7 milch cows, 1 other cattle, 7 calves dropped, 1 cattle purchased, 1 cattle sold living, 1 cattle slaughtered, 700 lbs. butter, 6 swine, 60 poultry in barnyard, 500 dozen eggs produced.
• 7 acres Indian corn, 300 bu. Indian corn, 7 acres oats, 300 bu. oats, 3/4 acre rye, 10 bu. rye, 8 acres wheat, 160 bu. wheat, 1/2 acre Irish potatoes, 60 bu. Irish potatoes, 1/4 acre tobacco, 700 lbs. tobacco.
• 3 acres with 20 apple trees, 15 bu. apples, 6 peach trees, 11 bu. peaches; total value of orchard products, $8.
• Market gardens valued at $50.
• 10 cords of wood cut valued at $10.

*Federal census lists lists Gerson, 25; Lizzie G., 25 (wife); Barton U., 4; Anna C., 7/12 (Nov.).*

### J. M. Kessler
• Owner
• 40 tilled acres, 5 acres permanent meadows, 16 acres unimproved land, valued at $4000.
• Farm machinery valued at $600; livestock, $600; fences, $50; fertilizers, $50; wages paid for farm labor, $100; weeks of hired labor, 20; estimated value of all farm productions, $600.

- Acres of grass mown, 18; not mown, 3; 20 tons of hay harvested.
- 4 horses, 2 mules and asses, 5 milch cows, 1 other cattle, 5 calves dropped, 1 cattle purchased, 2 cattle sold living, 500 lbs. butter, 6 swine, 50 poultry in barnyard, 400 dozen eggs produced.
- 5 acres Indian corn, 125 bu. Indian corn, 6 acres oats, 175 bu. oats, 1 acre rye, 10 bu. rye, 5 acres wheat, 100 bu. wheat, 1/2 acre Irish potatoes, 75 bu. Irish potatoes.
- 1/2 acre with 30 apple trees.
- 5 cords of wood cut valued at $5.

*Federal census lists J. M., 35; Susanna, 25; Elizabeth, 30 (sister); Henry F. Trostle, 20, boarder.*

## Henry Sensenig

- Owner
- 60 tilled acres, 5 acres permanent meadows, 10 acres woodland forests, 2 acres unimproved land, valued at $5000.
- Farm machinery valued at $600; livestock, $500; fences, $50; fertilizers, $50; wages paid for farm labor, $100; weeks of hired labor, 20; estimated value of all farm productions, $750.
- Acres of grass mown, 18; 20 tons of hay harvested.
- 3 horses, 5 milch cows, 1 other cattle, 5 calves dropped, 1 cattle purchased, 1 cattle sold living, 1 cattle slaughtered, 500 lbs. butter, 6 swine, 50 poultry in barnyard, 400 dozen eggs produced.
- 10 acres Indian corn, 300 bu. Indian corn, 10 acres oats, 300 bu. oats, 1 acre rye, 10 bu. rye, 10 acres wheat, 200 bu. wheat, 1/2 acre Irish potatoes, 50 bu. Irish potatoes.
- 1/2 acre with 25 apple trees.
- 5 cords of wood cut valued at $5.

*Federal census lists Henry, 42 (widower); Samuel, 7; Joseph, 5; Mary A., 6; Lydia, 4; Henry, 2; Rebecca Ream, 50, servant.*

## John Kramer

- Rents for shares of products.
- 60 tilled acres, 7 acres permanent meadows, 5 acres woodland forests, 2 acres unimproved land, valued at $4000.
- Farm machinery valued at $600; livestock, $600; fences, $100; fertilizers, $100; wages paid for farm labor, $100; weeks of hired labor, 20; estimated value of all farm productions, $600.
- Acres of grass mown, 14; not mown, 5; 15 tons of hay harvested.
- 3 horses, 9 milch cows, 4 other cattle, 9 calves dropped, 2 cattle sold living, 1 cattle slaughtered, 200 lbs. butter, 5 swine, 60 poultry in barnyard, 400 dozen eggs produced.
- 12 acres Indian corn, 250 bu. Indian corn, 7 acres oats, 180 bu. oats, 1/2 acre rye, 5 bu. rye, 12 acres wheat, 200 bu. wheat, 1/2 acre Irish potatoes, 80 bu. Irish potatoes. 1/16 acre tobacco, 25 lbs. tobacco.
- 1/2 acre with 20 trees, 8 peach trees.
- 15 cords of wood cut valued at $15.

*Federal census lists John, 28; Catherine, 28 (wife); Catherine Flick, 5 (stepdaughter); Laura, 2; Howard, 4/12 (Feb.); Lizzie Frakenbiller, 16,servant.*

## David Kessler

- Rents for shares of products.
- 35 tilled acres, 3 acres permanent meadows, 7 acres woodland acres, 6 acres permanent meadows, valued at $3000.
- Farm machinery valued at $300, livestock, $500; fences, $100; fertilizers, $100; wages paid for farm labor, $200; weeks of hired labor, 25; estimated value of all farm productions, $400.
- Acres of grass mown, 15; not mown, 3; 15 tons of hay harvested.
- 3 horses, 5 milch cows, 2 other cattle, 5 calves dropped, 1 cattle slaughtered, 400 lbs. butter, 5 swine, 40 poultry in barnyard, 200 dozen eggs produced.
- 5 acres Indian corn, 50 bu. Indian corn, 5 acres oats, 100 bu. oats, 2 acres rye, 10 bu. rye, 5 acres wheat, 100 bu. wheat, 1/2 acre Irish potatoes, 25 bu. Irish potatoes.
- 3 apple trees, 3 peach trees.
- 10 cords of wood cut valued at $10.

*Federal census lists David as Isaac, 66; Catherine, 67; Isaac, 23; Daniel, 25; Amanda, 22 (daughter-in-law); Harrison, 3; Catherine, 2; Joseph, 3/12 (Apr.); Lizzie Stover, 20, servant.*

## George Heft

- Rents for shares of products.
- 50 tilled acres, 4 acres permanent meadows, 30 acres woodland forests, 2 acres unimproved land, valued at $3000.
- Farm machinery valued at $600; livestock, $600; fences, $100; fertilizers, $100; wages paid for farm labor, $100; weeks of hired labor, 25; estimated value of all farm productions, $500.
- Acres of grass mown, 15; not mown, 10; 15 tons of hay harvested.
- 4 horses, 6 milch cows, 5 other cattle, 5 calves dropped, 1 cattle sold living, 1 cattle slaughtered, 500 lbs. butter, 6 swine, 50 poultry in barnyard, 300 dozen eggs produced.

- 8 acres Indian corn, 100 bu. Indian corn, 8 acres oats, 100 bu. oats, 2 acres rye, 20 bu. rye, 11 acres wheat, 200 bu. wheat, 1/2 acre Irish potatoes, 10 bu. Irish potatoes.
- 1/2 acre with 20 apple trees, 2 peach trees.

*Federal census lists George, 39; Mary, 39 (wife); Wayne, 18; Moses, 17; Samuel, 15; Charles, 12; Maggie, 8; John, 7; George, 2/12 (May). Also listed with George are Benjamin Kessler, 76, and wife Susan, 71.*

## William Kessler

- Owner
- 20 tilled acres, 1 acre permanent meadows, 2 acres woodland forests, 5 acres unimproved land, valued at $1000.
- Farm machinery valued at $200; livestock, $200; fences, $50; wages paid for farm labor, $25; weeks of hired labor, 4; estimated value of all farm productions, $100.
- Acres of grass mown, 3; not mown, 1; 3 tons of hay produced.
- 2 horses, 3 milch cows, 3 other cattle, 100 lbs. butter, 2 swine, 16 poultry in barnyard, 50 dozen eggs produced.
- 1 acre Indian corn, 20 bu. Indian corn, 18 acres oats, 50 bu. oats, 1/2 acre rye, 15 bu. rye, 2 acres wheat, 40 bu. wheat, 1/4 acre Irish potatoes, 10 bu. Irish potatoes.
- 4 apple trees.

*Federal census lists William, 49; Sophia, 43 (wife); Asop, 9.*

## John Gring

- Owner
- 20 tilled acres, 8 acres permanent meadows, 30 acres woodland forests, 10 acres unimproved land, valued at 2000.
- Farm machinery valued at $200; livestock, $400; fences, $50; fertilizers, $40; wages paid for farm labor, $20; weeks of hired labor, 4; estimated value of all farm productions, $500.
- Acres of grass mown, 14; now mown, 10; 15 tons of hay produced.
- 2 horses, 8 milch cows, 2 other cattle, 1 calves dropped, 1 cattle slaughtered, 800 lbs. butter, 3 swine, 50 poultry in barnyard, 250 dozen eggs produced, 4 acres Indian corn, 80 bu. Indian corn, 3 acres oats, 120 bu. oats, 1/2 acre rye, 5 bu. rye, 3 acres wheat, 60 bu. wheat, 1 acre Irish potatoes, 25 bu. Irish potatoes.
- 1/2 acre with 12 apple trees; 1 acre with 25 peach trees.
- Market gardens valued at $10.
- 15 cords of wood cut valued at $15.

*Federal census lists John, 36; Catherine Ann, 39 (wife); John, 10; Samuel, 8; Charles, 6; Mary E., 4; Catherine, 2; Franklin, 2; Elizabeth, 68 (mother).*

## John Overholtzer

- Owner
- 75 tilled acres, 12 acres permanent meadows, 16 acres woodland forests, 15 acres unimproved land, valued at $6000.
- Farm machinery valued at $600; livestock, $700; fences, $100; fertilizers, $100; wages paid for farm labor, $150; weeks of hired labor, 12; estimated value of all farm productions, $1000.
- Acres of grass mown, 28; not mown, 10; 40 tons of hay harvested.
- 4 horses, 9 milch cows, 4 other cattle, 9 calves dropped, 3 cattle sold living, 1 cattle slaughtered, 1000 lbs. butter, 8 swine, 100 poultry in barnyard, 700 dozen eggs produced.
- 12 acres Indian corn, 400 bu. Indian corn, 11 acres oats, 400 bu. oats, 1/2 acre rye, 25 bu. rye, 12 acres wheat, 200 bu. wheat, 1 acre Irish potatoes, 75 bu. Irish potatoes.
- 1/2 acre with 50 apple trees, 20 bu. apples; 1 acre with 100 peach trees; total value of orchard products, $25.
- 45 cords of wood cut valued at $45.

*Federal census lists John, 33; Elizabeth, 29 (wife); Agnes, 6.*

## Samuel Burkhard

- Owner
- 20 tilled acres, valued at $1500.
- Farm machinery valued at $200; livestock, $200; fences, $25; fertilizers, $25; wages paid for farm labor, $25; weeks of hired labor, 4; estimated value of all farm productions, $200.
- Acres of grass mown, 7; not mown, 7; 6 tons of hay harvested.
- 4 milch cows, 4 calves dropped, 300 lbs. butter, 2 swine, 30 poultry in barnyard, 200 dozen eggs produced.
- 1 acre Indian corn, 20 bu. Indian corn, 1 acre oats, 30 bu. oats, 1/2 acre rye, 10 bu. rye, 2 acres wheat, 48 bu. wheat, 1/2 acre Irish potatoes, 25 bu. Irish potatoes.
- 1/2 acre with 12 apple trees.

*Federal census lists Samuel, 53, blacksmith; Sophia, 48 (wife); Charles, 16; Catherine, 14; Mary, 12; Sally, 9; Samuel, 8; Andy, 1; Augustine, 23, works on farm; Isaac Bordenhart, 38, stone mason; Lizzie, 39 (wife).*

## John Fries

- Owner
- 100 tilled acres, 7 acres permanent meadows, 12 acres woodland forests, 60 acres unimproved land, valued at $6000.
- Farm machinery valued at $1000; livestock, $1000; fences, $150; fertilizers, $100; wages paid for farm labor, $200; weeks of hired labor, 40; estimated value of all farm productions, $900.
- Acres of grass mown, 25; not mown, 20; 25 tons of hay harvested.
- 2 horses, 7 milch cows, 3 other cattle, 7 calves dropped, 2 cattle sold living, 1 cattle slaughtered, 450 lbs. butter, 4 swine, 40 poultry in barnyard, 450 dozen eggs produced.
- 12 acres Indian corn, 200 bu. Indian corn, 10 acres oats, 200 bu. oats, 3 acres rye, 12 bu. rye, 12 acres wheat, 175 bu. wheat, 1/2 acre Irish potatoes, 75 bu. Irish potatoes.
- 1/2 acre with 20 apple trees; 1 acre with 200 peach trees; 125 bu. peaches; total value of orchard products, $30.
- Market gardens valued at $25.
- 10 cords of wood cut valued at $10.

*Federal census lists John, 64; Mary Ann, 62 (wife); Hannah, 27; Sarah, 25; Israel, 22; Emma, 20; John Hoch, 2 (grandson); Elizabeth Folk, 1 (granddaughter).*

## Christian Messner

- Rents for shares of products.
- 40 tilled acres, valued at $2000.
- Farm machinery valued at $200; livestock, $200; fences, $50; fertilizers, $50; wages paid for farm labor, $50; weeks of hired labor, 6; estimated value of all farm productions, $300.
- Acres of grass mown, 12; 12 tons of hay harvested.
- 2 horses, 4 milch cows, 4 calves dropped, 300 lbs. butter, 3 swine; 30 poultry in barnyard, 100 dozen eggs produced.
- 8 acres Indian corn, 100 bu. Indian corn, 6 acres oats, 200 bu. oats, 1 acre rye, 10 bu. rye, 8 acres wheat, 130 bu. wheat, 1/2 acre Irish potatoes, 20 bu. Irish potatoes.
- 1/2 acre with 20 apple trees; 5 peach trees.

*Federal census lists Christian, 39; Rebecca, 40 (wife); John, 11; William Zerbe, 23; Ellen, 24 (wife); Harvey, 1.*

## John Bachman

- Rents for shares of products.
- 20 tilled acres, valued at $1000.
- Farm machinery valued at $200; livestock, $200; fences, $50; fertilizers, $25; estimated value of all farm productions, $300.
- Acres of grass mown, 6; 6 tons of hay harvested.
- 1 horse, 2 milch cows, 2 other cattle, 2 calves dropped, 100 lbs. butter, 3 swine, 30 poultry in barnyard, 100 dozen eggs produced.
- 4 acres Indian corn, 100 bu. Indian corn, 4 acres oats, 80 bu. oats, 1 acre rye, 10 bu. rye, 4 acres wheat, 50 bu. wheat, 1/2 acre Irish potatoes, 20 bu. Irish potatoes.
- 1/2 acre with 40 apple trees; 1/4 acre with 10 peach trees.

*Federal census lists no John Bachman.*

## Edward Lorah

- Owner
- 25 tilled acres, valued at $1400.
- Farm machinery valued at $200; livestock, $200; fences, $50; fertilizers, $25; wages paid for farm labor, $25; estimated value of all farm productions, $400.
- Acres of grass mown, 7; 7 tons of hay harvested.
- 2 horses, 4 milch cows, 1 other cattle, 4 calves dropped; 200 lbs. butter, 4 swine, 25 poultry in barnyard, 100 dozen eggs produced.
- 4 acres Indian corn, 75 bu. Indian corn, 5 acres oats, 100 bu. oats, 1 acre rye, 1 bu. rye, 3 acres wheat, 60 bu. wheat, 1/2 acre Irish potatoes, 26 bu. Irish potatoes.
- 1/2 acre with 10 apple trees; 1/2 acre with 6 peach trees.
- 5 cords of wood cut valued at $5.

*Federal census lists Edward, 47; Amanda, 40 (wife); Rachel, 19; Francis, 13; Adam, 11; John, 8; Charles, 8; Nancy, 3; Harry, 1.*

## Isaac Weis

- Owner
- 50 tilled acres, valued at $1000.
- Farm machinery valued at $200; livestock, $200; fences, $50; fertilizers, $25; wages paid for farm labor, $25; estimated value of all farm productions, $400.
- Acres of grass mown, 14; 8 tons of hay harvested.

- 2 horses, 2 milch cows, 2 calves dropped, 200 lbs. butter, 2 swine, 15 poultry in barnyard, 100 dozen eggs produced.
- 5 acres Indian corn, 75 bu. Indian corn, 4 acres oats, 75 bu. oats, 1 acre rye, 12 bu. rye, 4 acres wheat, 80 bu. wheat, 1/2 acre Irish potatoes, 30 bu. Irish potatoes.
- 1/4 acre with 20 apple trees.
- 8 cords of wood cut valued at $8.

*Federal census lists Isaac, 30; Mary Ann, 28 (wife); John, 9; Nicholas, 6; John Weis, 73 (father); Catherine, 69 (wife); Betsy, 69 (sister); Eli Griffy, 40 (laborer).*

### Isais Burkholder

- Owner
- 8 tilled acres, 3 acres permanent meadows, 5 acres woodland forests, 5 acres unimproved land, valued at $2000.
- Farm machinery valued at $200; livestock, $200; fences, $50; fertilizers, $25; wages paid for farm labor, $10; weeks of hired labor, 2; estimated value of all farm productions, $200.
- Acres of grass mown, 6; not mown, 1; 10 tons of hay harvested.
- 1 horse, 3 milch cows, 1 other cattle, 3 calves dropped, 1 cattle sold living; 300 lbs. butter, 4 swine, 9 poultry in barnyard, 50 dozen eggs produced.
- 3 acres Indian corn, 100 bu. Indian corn, 3 acres oats, 110 bu. oats, 3 acres wheat, 50 bu. wheat, 1/4 acre Irish potatoes, 25 bu. Irish potatoes.
- 1/4 acre with 25 apple trees, 25 bu. apples.
- 6 cords of wood cut valued at $6.

*Federal census lists Isais, 50; Sarah, 43 (wife); William, 16; Mary, 8; Nancy, 82 (stepmother).*

### Menno Good

- Owner
- 35 tilled acres, 3 acres permanent meadows, 4 acres woodland forests, 1 acre unimproved land, valued at $3000.
- Farm machinery valued at $200; livestock, $500; fences, $50; fertilizers, $40; wages paid for farm labor, $25; weeks of hired labor, 4; estimated value of all farm productions, $500.
- Acres of grass mown, 11; not mown, 4; 6 tons of hay harvested, 3 bu. clover seed.
- 2 horses, 4 milch cows, 1 other cattle, 4 calves dropped, 400 lbs. butter, 3 swine, 20 poultry in barnyard, 3 other poultry, 100 dozen eggs produced.
- 7 acres Indian corn, 60 bu. Indian corn, 6 acres oats, 150 bu. oats, 5 acres wheat, 60 bu. wheat, 1/4 acre Irish potatoes, 10 bu. Irish potatoes.
- 1/4 acre with 3 apple trees; 1/4 acre with 12 peach trees; 6 bu. peaches; total value of orchard products, $6.

*Federal census lists Menno, 24; Lizzie, 23 (wife); Mary, 3/12 (Feb.); Judith, 48 (widowed mother); Anna, 20 (sister).*

### Amos Good

- Owner
- 50 tilled acres, 10 acres permanent meadows, 17 acres woodland forests, 25 acres unimproved land, valued at $4000.
- Farm machinery valued at $400; livestock, $500; fences, $50; fertilizers, $75; wages paid for farm labor, $75; weeks of hired labor, 12; estimated value of all farm productions, $680.
- Acres of grass mown, 18; not mown, 5; 20 tons of hay harvested, 5 bu. clover seed; 2 bu. grass seed.
- 3 horses, 5 milch cows, 8 other cattle, 5 calves dropped, 2 cattle sold living, 1 cattle slaughtered, 600 lbs. butter, 5 swine, 20 poultry in barnyard, 100 dozen eggs produced.
- 9 acres Indian corn, 200 bu. Indian corn, 8 acres oats, 250 bu. oats, 1 acre rye, 8 bu. rye, 8 acres wheat, 140 bu. wheat, 1 acre Irish potatoes, 20 bu. Irish potatoes, 2 bu. sweet potatoes, 1/4 acre tobacco, 700 lbs. tobacco.
- 1/2 acre with 20 apple trees; 12 peach trees, 4 bu. peaches; total value of orchard products, $4.
- 6 cords of wood cut valued at $6.

*Federal census lists Amos, 44; Lavinia, 39 (wife); Eli, 17; Sybill, 15; Lizzie, 11; Catharine, 8; Hettie A., 4; Esther, 72 (mother).*

### William Good

- Owner
- 46 tilled acres, 5 acres permanent meadows, 6 acres woodland forests, valued at $4000.
- Farm machinery valued at $200; livestock, $300; fences, $25; fertilizers, $25; wages paid for farm labor, $25; weeks of hired labor, 4; estimated value of all farm productions, $400.
- Acres of grass mown, 12; not mown, 5; 12 tons of hay harvested, 1/2 bu. clover seed, 1/2 bu. grass seed.
- 2 horses, 5 milch cows, 5 other cattle, 5 calves dropped, 1 cattle sold living, 1 cattle slaughtered, 300 lbs. butter, 4 swine, 50 poultry in barnyard, 200 dozen eggs produced.
- 5 acres Indian corn, 100 bu. Indian corn, 4 acres oats, 150 bu. oats, 1/2 acre rye, 7 bu. rye, 5 acres wheat, 80 bu. wheat, 1/4 acre Irish potatoes, 15 bu. Irish potatoes.

• 2 acres with 40 apple trees, 25 bu. apples; total value of orchard products, $10.
• 4 cords of wood cut valued at $4.
*Federal census lists William, 35; Catherine, 36 (wife).*

## Samuel Trostle
• Rents for shares of products.
• 25 tilled acres, 5 acres permanent meadows, 9 acres unimproved land, valued at $1500.
• Farm machinery valued at $200; livestock, $200; fences, $25; fertilizers, $50; estimated value of all farm productions, $200.
• Acres of grass mown, 7; not mown, 1; 7 tons of hay harvested.
• 2 horses, 4 milch cows, 4 calves dropped, 1 cattle slaughtered, 300 lbs. butter, 2 swine, 20 poultry in barnyard, 100 dozen eggs produced.
• 3 acres Indian corn, 75 bu. Indian corn, 3 acres oats, 75 bu. oats, 3 acres wheat, 60 bu. wheat, 1/4 acre Irish potatoes, 15 bu. Irish potatoes.
• 1/2 acre with 20 apple trees, 5 bu. apples.
*Federal census lists Samuel, 31; Susanna, 29 (wife); John, 7; William, 3; Harry, 2; John Zerbe, 59, farmer; Susanna, 56 (wife).*

## Samuel Musser
• Owner
• 50 tilled acres, 2 acres permanent meadows, 30 acres woodland forests, 5 acres unimproved land, valued at $4000.
• Farm machinery valued at $600; livestock, $600; fences, $50; fertilizers, $50; wages paid for farm labor, $100; weeks of hired labor, 20; estimated value of all farm productions, $600.
• Acres of grass mown, 10; not mown, 10; 20 tons of hay harvested, 3 bu. clover seed, 1 bu. grass seed.
• 2 horses, 1 mules and asses, 5 milch cows, 5 other cattle, 5 calves dropped, 1 cattle slaughtered, 500 lbs. butter, 4 swine, 40 poultry in barnyard, 800 dozen eggs produced.
• 8 acres Indian corn, 150 bu. Indian corn, 8 acres oats, 250 bu. oats, 3 acres rye, 25 bu. rye, 8 bu. wheat, 160 bu. wheat, 1/2 acre Irish potatoes, 50 bu. Irish potatoes.
• 2 acres with 60 apple trees, 50 bu. apples; 1/2 acre with 25 peach trees, 4 bu. peaches; total value of orchard products, $20.
• 10 cords of wood cut valued at $10.
*Federal census lists Samuel, 31; Mary, 26 (wife); Susanna, 5; Henry, 4; Israel, 2; Lizzie, 11/12 (July); Elizabeth, 66 (mother); Lizzie Kohl, 24, servant.*

## Samuel Kieffer
• Owner
• 50 tilled acres, 2 acres permanent meadows, 30 acres woodland forests, 5 acres unimproved land, valued at $4000.
• Farm machinery valued at $600; livestock, $600; fences, $50; fertilizers, $50; wages paid for farm labor, $100; weeks of hired labor, 20; estimated value of all farm productions, $770.
• Acres of grass mown, 10; not mown, 10; 20 tons of hay harvested, 3 bu. clover seed, 1 bu. grass seed.
• 3 horses, 4 milch cows, 4 other cattle, 4 calves dropped, 1 cattle slaughtered, 400 lbs. butter, 4 swine, 40 poultry in barnyard, 600 dozen eggs produced.
• 8 acres Indian corn, 150 bu. Indian corn, 8 acres oats, 250 bu. oats, 3 acres rye, 25 bu. rye, 8 acres wheat, 260 bu. wheat, 1/2 acre Irish potatoes, 50 bu. Irish potatoes.
• 2 acres with 50 apple trees, 25 bu. apples; 1/2 acre with 20 peach trees, 2 bu. peaches; total value of orchard products, $20.
• 10 cords of wood cut valued at $10.
*Federal census lists Samuel, 60; Susanna, 60 (wife); Elizabeth, 20; Solomon Coldren, 1 (grandchild).*

## Eli Kraemer
• Owner
• 50 tilled acres, 2 acres permanent meadows, 10 acres woodland forests, 5 acres unimproved land, valued at $2000.
• Farm machinery valued at $600; livestock, $600; fences, $50; fertilizers, $50; wages paid for farm labor, $100; weeks of hired labor, 20; estimated value of all farm productions, $600.
• Acres of grass mown, 10; not mown, 10; 15 tons of hay harvested.
• 2 horses, 4 milch cows, 4 other cattle, 4 calves dropped, 1 cattle slaughtered, 400 lbs. butter, 4 swine, 50 poultry in barnyard, 300 dozen eggs produced.
• 8 acres Indian corn, 100 bu. Indian corn, 5 acres oats, 100 bu. oats, 4 acres wheat, 100 bu. wheat, 1/2 acre Irish potatoes, 25 bu. Irish potatoes.
• 1 acre with 50 apple trees, 10 bu. apples; 20 peach trees.
• 10 cords of wood cut valued at $10.
*Federal census lists Eli, 51; Leah, 42 (wife); Mary E., 21; Leah E., 18; Christian, 15; Hannah A., 13; Isabella H., 6; William, 2.*

### Abraham Gehman

- Owner
- 60 tilled acres, 8 acres permanent meadows, 6 acres woodland forests, 15 acres unimproved land, valued at $6000.
- Farm machinery valued at $600; livestock, $600; fences, $50; fertilizers, $60; wages paid for farm labor, $100; weeks of hired labor, 40; estimated value of all farm productions, $600.
- Acres of grass mown, 8; not mown, 8; 16 tons of hay harvested, 2 bu. clover seed, 1 bu. grass seed.
- 3 horses, 4 milch cows, 4 other cattle, 4 calves dropped, 1 cattle sold living, 1 cattle slaughtered, 400 lbs. butter produced, 2 swine, 50 poultry in barnyard, 400 dozen eggs produced.
- 9 acres Indian corn, 400 bu. Indian corn, 12 acres oats, 300 bu. oats, 3 acres rye, 40 bu. rye, 7 acres wheat, 140 bu. wheat, 1/2 acre Irish potatoes, 100 bu. Irish potatoes.
- 1 acre with 50 apple trees, 20 bu. apples; total value of orchard products, $10.
- 10 cords of wood cut valued at $10.

*Federal census lists Abraham, 32; Elizabeth, 30 (wife); Israel, 7; Susanna, 2; Jonathan, 10/12 (Aug.); Joshua, 33 (brother).*

### Schnader & Good

- Owner
- 25 tilled acres, 3 acres permanent meadows, 2 acres woodland forests, 3 acres unimproved land, valued at $3000.
- Farm machinery valued at $300; livestock, $600; fences, $50; fertilizers, $50; wages paid for farm labor, $10; weeks of hired labor, 2; estimated value of all farm productions, $500.
- Acres of grass mown, 12; 10 tons of hay harvested.
- 4 horses, 3 milch cows, 3 other cattle, 3 calves dropped, 1 cattle slaughtered, 300 lbs. butter, 5 swine, 50 poultry in barnyard, 300 dozen eggs produced.
- 6 acres Indian corn, 100 bu. Indian corn, 5 acres oats, 170 bu. oats, 1 acre rye, 12 bu. rye, 5 acres wheat, 75 bu. wheat, 1/2 acre Irish potatoes, 100 bu. Irish potatoes.
- 1 acre with 40 apple trees, 10 peach trees.
- 8 cords of wood cut valued at $8.

*Federal census lists John Schnader, 47; Polly, 46 (wife); Mary, 23; John, 20; Benjamin Good, 76 (father-in-law); Richard Good, 35 (brother-in-law).*

### Reuben Slabach

- Owner
- 30 tilled acres, 3 acres permanent meadows, 6 acres woodland forests, 10 acres unimproved land, valued at $3000.
- Farm machinery valued at $300; livestock, $600; fences, $50; fertilizers, $50; wages paid for farm labor, $25; weeks of hired labor, 4; estimated value of all farm productions, $500.
- Acres of grass mown, 12, 15 tons of hay harvested, 1 bu. clover seed, 1 bu. grass seed.
- 2 horses, 3 milch cows, 6 other cattle, 3 calves dropped, 1 cattle sold living, 1 cattle slaughtered, 300 lbs. butter, 5 swine, 50 poultry in barnyard, 300 dozen eggs produced.
- 6 acres Indian corn, 150 bu. Indian corn, 5 acres oats, 200 bu. oats, 2 acres rye, 20 bu. rye, 6 acres wheat, 100 bu. wheat, 1 acre Irish potatoes, 150 bu. Irish potatoes.
- 2 acres with 12 apple trees; 1/4 acre with 10 peach trees.
- 10 cords of wood cut valued at $10.

*Federal census lists Reuben, 26; Catherine, 26 (wife); Harriet, 5; Anna, 3; George Shupp, 56 (cooper and Reuben's uncle).*

### Isaac Shupp

- Owner
- 30 tilled acres, valued at $1000.
- Farm machinery valued at $200; livestock, $300; fences, $50; fertilizers, $50; wages paid for farm labor, $25; weeks of hired labor, 4; estimated value of all farm productions, $200.
- Acres of grass mown, 15; not mown, 10; 15 tons of hay harvested, 2 bu. clover seed.
- 2 horses, 3 milch cows, 2 other cattle, 3 calves dropped, 1 cattle slaughtered, 300 lbs. butter, 3 swine, 50 poultry in barnyard, 1 other poultry, 300 dozen eggs produced.
- 5 acres Indian corn, 50 bu. Indian corn, 4 acres oats, 75 bu. oats, 4 acres wheat, 40 bu. wheat, 1/2 acre Irish potatoes, 30 bu. Irish potatoes.
- 1 acre with 20 apple trees.
- 10 cords of wood cut valued at $10.

*Federal census lists Isaac, 70; Harriet, 70 (wife); Reuben, 23; Anna, 23 (wife); Anna, 3.*

### Charles Bicher

- Owner
- 30 tilled acres, 1 acre permanent meadows, 1 acre woodland forests, 18 acres unimproved land, valued at $2000.

- Farm machinery valued at $200; livestock, $300; fences, $25; fertilizers, $25; wages paid for farm labor, $25; weeks of hired labor, 6; estimated value of all farm productions, $250.
- Acres of grass mown, 10; 10 tons of hay harvested.
- 3 horses, 2 milch cows, 5 other cattle, 2 calves dropped, 1 cattle slaughtered, 200 lbs. butter, 4 swine, 20 poultry in barnyard, 100 dozen eggs produced.
- 4 acres Indian corn, 100 bu. Indian corn, 4 acres oats, 100 bu. oats, 1/4 acre rye, 3 bu. rye, 4 acres wheat, 48 bu. wheat, 1/2 acre Irish potatoes, 50 bu. Irish potatoes.
- 1/4 acre with 15 apple trees.
- 5 cords of wood cut valued at $5.

*Federal census lists Charles, 36; Lizzie, 37 (wife); Mary Ann, 9; James, 9.*

## Isaac Roth

- Owner
- 60 tilled acres, 9 acres permanent meadows, 20 acres woodland forests, 60 acres unimproved land, valued at $3000.
- Farm machinery valued at $300; livestock, $600; fences, $100; fertilizers, $48; wages paid for farm labor, $30; weeks of hired labor, 5; estimated value of all farm productions, $600.
- Acres of grass mown, 20; not mown, 10; 25 tons of hay harvested, 2 bu. clover seed, 3 bu. grass seed.
- 3 horses, 5 milch cows, 6 other cattle, 5 calves dropped, 2 cattle sold living, 1 cattle slaughtered, 300 lbs. butter, 5 swine, 20 poultry in barnyard, 150 dozen eggs produced.
- 10 acres Indian corn, 400 bu. Indian corn, 10 acres oats, 225 bu. oats, 1 acre rye, 10 bu. rye, 8 acres wheat, 200 bu. wheat, 1 acre Irish potatoes, 150 bu. Irish potatoes, 1/8 acre tobacco, 100 lbs. tobacco.
- 5 acres with 100 apple trees, 10 bu. apples, 25 peach trees.
- 15 cords of wood cut valued at $15.

*Federal census lists Isaac, 66; Fianna, 53 (wife); Elizabeth, 21; Lydia, 19; Adeline, 17; Frank, 35; David, 16; Isaac, 12; Jacob, 14; Edward, 8.*

## Peter Becker

- Owner
- 70 tilled acres, 6 acres permanent meadows, 2 acres woodland forests, 2 acres unimproved land, valued at $5000.
- Farm machinery valued at $500; livestock, $800; fences, $200; fertilizers, $200; wages paid for farm labor, $80; weeks of hired labor, 8; estimated value of all farm productions, $800.
- Acres of grass mown, 20; not mown, 10; 25 tons of hay harvested.
- 5 horses, 7 milch cows, 8 other cattle, 7 calves dropped, 1 cattle slaughtered, 700 lbs. butter, 4 swine, 60 poultry in barnyard, 500 dozen eggs produced.
- 14 acres Indian corn, 300 bu. Indian corn, 12 acres oats, 300 bu. oats, 12 acres wheat, 300 bu. wheat, 1 acre Irish potatoes, 60 bu. Irish potatoes.
- 1 acre with 50 apple trees; 1/4 acre with 60 peach trees.
- 8 cords of wood cut valued at $8.

*Federal census lists Peter, 46; Hannah, 41 (wife); John, 18; Amelia, 16; Lizzie, 11; Peter, 5.*

## John Gehman

- Owner
- 50 tilled acres, 5 acres permanent meadows, 2 acres woodland forests, 2 acres unimproved land, valued at $3000.
- Farm machinery valued at $500; livestock, $500; fences, $50; fertilizers, $50; wages paid for farm labor, $125; weeks of hired labor, 20; estimated value of all farm productions, $800.
- Acres of grass mown, 20; not mown, 10; 25 tons of hay harvested.
- 5 horses, 4 milch cows, 4 calves dropped, 400 lbs. butter, 3 swine, 50 poultry in barnyard, 300 dozen eggs produced.
- 10 acres Indian corn, 250 bu. Indian corn, 10 acres oats, 250 bu. oats, 10 acres wheat, 250 bu. wheat, 1 acre Irish potatoes, 50 bu. Irish potatoes.
- 1 acre with 50 apple trees, 10 bu. apples; 1/4 acre with 25 peach trees, 5 bu. peaches; total value of orchard products, $10.
- 5 cords of wood cut valued at $5.

*Federal census lists John, 69; Judith, 61 (wife); Anna, 21; Josiah Kurtz, 16, servant.*

## John Hershberger

- Owner
- 100 tilled acres, 4 acres permanent meadows, 6 acres woodland forests, 2 acres unimproved land, valued at $7000.
- Farm machinery valued at $600; livestock, $600; fences, $100; fertilizers, $100; wages paid for farm labor, $150; weeks of hired labor, 40; estimated value of all farm productions, $1200.

- Acres of grass mown, 25; not mown, 10; 30 tons of hay harvested.
- 4 mules and asses, 8 milch cows, 10 other cattle, 8 calves dropped, 2 cattle sold living, 1 cattle slaughtered, 500 lbs. butter, 8 sheep, 1 sheep sold living, 8 fleeces, 20 lbs. wool, ? swine, 50 poultry in barnyard, 300 dozen eggs produced.
- 12 acres Indian corn, 600 bu. Indian corn, 12 acres oats, 500 bu. oats, 2 acres rye, 20 bu. rye, 20 acres wheat, 350 bu. wheat, 2 acres Irish potatoes, 100 bu. Irish potatoes.
- 2 acres with 70 apple trees; 1/4 acre with 10 peach trees, 5 bu. peaches; total value of orchard products, $5.
- 5 cords of wood cut valued at $5.

*Federal census lists Catherine Glass, 66 (widow); John Glass, 22, farm laborer; John Hershberger, 25 (son-in-law), farm laborer; Susan Hershberger, 25 (daughter); Jesse Ann, 6 (granddaughter); Henry, 3 (grandson).*

## Henry Bauman

- Owner
- 18 tilled acres, 2 acres permanent meadows, 2 acres woodland forests, 1 acre unimproved land, valued at $1900.
- Farm machinery valued at $150; livestock, $200; fences, $50; fertilizers, $20; wages paid for farm labor, $25; estimated value of all farm productions, $400.
- Acres of grass mown, 10; 15 tons of hay harvested.
- 1 horse, 2 milch cows, 1 other cattle, 2 calves dropped, 200 lbs. butter, 2 swine, 20 poultry in barnyard, 250 dozen eggs produced.
- 3 acres Indian corn, 100 bu. Indian corn, 3 acres oats, 120 bu. oats, 3 acres wheat, 60 bu. wheat, 1/2 acre Irish potatoes, 30 bu. Irish potatoes, 2 bu. sweet potatoes.
- 1 acre with 10 apple trees; 10 peach trees; total value of orchard products, $5.
- 2 cords of wood cut valued at $2.

*Federal census lists Henry, 40; Matty, 40 (wife).*

## Benjamin Good

- Owner
- 15 tilled acres, 1/2 acre permanent meadows, 3 acres woodland forests, 1 acre unimproved land, valued at $1000.
- Farm machinery valued at $200; livestock, $200; fertilizers, $25; wages paid for farm labor, $25; weeks of hired labor, 4; estimated value of all farm productions, $300.
- Acres of grass mown, 4; 5 tons of hay harvested.
- 1 horse, 2 milch cows, 1 other cattle, 2 calves dropped, 150 lbs. butter, 3 swine, 30 poultry in barnyard, 200 dozen eggs produced.
- 4 acres Indian corn, 60 bu. Indian corn, 2 acres wheat, 30 bu. wheat, 1/4 acre Irish potatoes, 25 bu. Irish potatoes.
- 1 acre with 25 apple trees, 6 peach trees; total value of orchard products, $1.

*Federal census lists Benjamin, 49; Amanda, 36 (wife); James, 17.*

## Catharine Coldren

- Owner
- 30 tilled acres, valued at $2000.
- Farm machinery valued at $200; livestock, $200; fences, $25; fertilizers, $25; wages paid for farm labor, $25; weeks of hired labor, 4; estimated value of all farm productions, $300.
- Acres of grass mown, 7; 7 tons of hay harvested.
- 1 horse, 2 milch cows, 2 calves dropped, 2 cattle sold living, 200 lbs. butter, 2 swine, 30 poultry in barnyard, 200 dozen eggs produced.
- 3 acres Indian corn, 100 bu. Indian corn, 3 acres oats, 100 bu. oats, 4 acres wheat, 40 bu. wheat, 1/4 acre Irish potatoes, 25 bu. Irish potatoes.
- 1/2 acre with 25 apple trees; 12 peach trees; 1 bu. peaches; total value of orchard products, $1.
- 4 cords of wood cut valued at $4.

*Federal census lists Catharine, 72; Harriet, 45.*

## Samuel Stoner

- Owner
- 12 tilled acres, 1 acre permanent meadows, 2 acres woodland forests, 3 acres unimproved land, valued at $1000.
- Farm machinery valued at $200; livestock, $200; fences, $25; fertilizers, $25; wages paid for farm labor, $25; weeks of hired labor, 4; estimated value of all farm production, $300.
- Acres of grass mown, 2; 5 tons of hay harvested.
- 1 horse, 2 milch cows, 2 other cattle, 2 calves dropped, 150 lbs. butter, 2 swine, 20 poultry in barnyard, 100 dozen eggs produced.
- 1 acre Indian corn, 25 bu. Indian corn, 1 acre oats, 50 bu. oats, 2 acres wheat, 40 bu. wheat, 1/4 acre Irish potatoes, 10 bu. Irish potatoes.
- 1/2 acre with 12 apple trees, 6 peach trees.
- 2 cords of wood cut valued at $2.

*Federal census lists Samuel Steiner, 46, a stone mason; Elizabeth, 38 (wife); Margaret, 12; David, 10; Susanna, 67 (mother).*

### Peter Frankhauser

- Owner
- 50 tilled acres, 4 acres permanent meadows, 7 acres woodland forests, 5 acres unimproved land, valued at $2800.
- Farm machinery valued at $1500; livestock, $600; fences, $100; fertilizers, $60; wages paid for farm labor, $100; weeks of hired labor, 12; estimated value of all farm productions, $600.
- Acres of grass mown, 10; not mown, 10; 10 tons of hay harvested, 4 bu. clover seed, 1/2 bu. grass seed.
- 3 horses, 4 milch cows, 4 other cattle, 4 calves dropped, 400 lbs. butter, 4 swine, 2 poultry in barnyard, 2 other poultry, 300 dozen eggs produced.
- 10 acres Indian corn, 200 bu. Indian corn, 8 acres oats, 250 bu. oats, 4 acres rye, 30 bu. rye, 10 acres wheat, 100 bu. wheat, 1 acre Irish potatoes, 40 bu. Irish potatoes, 6 bu. sweet potatoes, 1 acre tobacco, 1430 lbs. tobacco.
- 3 acres with 50 apple trees, 10 bu. apples; 10 peach trees, 1 bu. peaches.
- 3 cords of wood cut valued at $3.

*Federal census lists Peter, 26; Lydia, 26 (wife); Monroe Slabach, 16 (foundling/orphan).*

### Henry Glass

- Rents for shares of products.
- 36 tilled acres, 5 acres permanent meadows, 6 acres woodland forests, 7 acres unimproved land, valued at $3000.
- Farm machinery valued at $150; livestock, $300; fences, $50; fertilizers, $36; wages paid for farm labor, $50; weeks of hired labor, 12; estimated value of all farm productions, $400.
- Acres of grass mown, 20; not mown, 10; 10 tons of hay harvested, 4 bu. clover seed.
- 2 horses, 5 milch cows, 4 other cattle, 4 calves dropped, 2 cattle purchased, 3 cattle sold living, 1 cattle slaughtered, 250 lbs. butter, 5 swine, 10 poultry in barnyard, 200 dozen eggs produced.
- 6 acres Indian corn, 100 bu. Indian corn, 5 acres oats, 150 bu. oats, 2 acres rye, 12 bu. rye, 8 acres wheat, 145 bu. wheat, 1/2 acre Irish potatoes, 10 bu. Irish potatoes, 2 bu. sweet potatoes.
- 2 acres with 12 apple trees; 6 peach trees, 2 bu. peaches; total value of orchard products, $1.
- 5 cords of wood cut valued at $5.

*Federal census lists Henry, 30; Anna, 23 (wife); Wilson S., 2; Catherine Good, 65 (mother-in-law); Catherine Zeam, 16 (servant).*

### Isaac Coldren

- Owner
- 40 tilled acres, 14 acres permanent meadows, 10 acres woodland forests, 5 acres unimproved land, valued at $2500.
- Farm machinery valued at $600; livestock, $500; fences, $200; fertilizers, $200; wages paid for farm labor, $500; weeks of hired labor, 50; estimated value of all farm productions, $500.
- Acres of grass mown, 18; not mown, 5; 15 tons of hay harvested.
- 4 horses, 4 milch cows, 2 other cattle, 4 calves dropped, 1 cattle purchased, 1 cattle slaughtered, 400 lbs. butter, 4 swine, 20 poultry in barnyard, 200 dozen eggs produced.
- 6 acres Indian corn, 100 bu. Indian corn, 6 acres oats, 200 bu. oats, 2 acres rye, 18 bu. rye, 6 acres wheat, 66 bu. wheat, 1/2 acre Irish potatoes, 25 bu. Irish potatoes.
- 1/2 acre with 40 apple trees.
- Market gardens valued at $2.
- 6 cords of wood cut valued at $6.

*Federal census lists Isaac, 52; Ellen, 40 (wife); Annie, 4; Lydia, 2.*

### Henry Weaver

- Owner
- 20 tilled acres, 2 acres permanent meadows, valued at $1000.
- Farm machinery valued at $100; livestock, $100; fences, $25, fertilizers, $25, wages paid for farm labor, $20; weeks of hired labor, 1; estimated value of all farm productions, $500.
- Acres of grass mown, 5; not mown, 1; 5 tons of hay harvested.
- 1 horse, 1 milch cow, 1 other cattle, 1 calf dropped, 200 lbs. butter, 3 swine, 14 poultry in barnyard, 200 dozen eggs produced.
- 1 acre Indian corn, 25 bu. Indian corn, 1 acre oats, 25 bu. oats, 1 acre rye, 5 bu. rye, 1 acre wheat, 20 bu. wheat, 1/2 acre Irish potatoes, 25 bu. Irish potatoes, 1/16 acre tobacco, 200 lbs. tobacco.
- 1/2 acre with 10 apple trees; 10 peach trees.
- 2 cords of wood cut valued at $2.

*Federal census lists Henry, 46; Susan, 42 (wife); Cyrus, 21; Eve Beck, 59 (sister-in-law).*

### William Weinhold

- Owner
- 25 tilled acres, 2 acres permanent meadows, 3 acres woodland forests, 2 acres unimproved land, valued at $1000.

- Farm machinery valued at $100; livestock, $100; fences, $25; fertilizers, $25; wages paid for farm labor, $20; weeks of hired labor, 1; estimated value of all farm productions, $500.
- Acres of grass mown, 3; not mown, 1; 5 tons of hay harvested.
- 1 horse, 2 milch cows, 2 other cattle, 2 calves dropped, 200 lbs. butter, 3 swine, 25 poultry in barnyard, 200 dozen eggs produced.
- 2 acres Indian corn, 50 bu. Indian corn, 2 acres oats, 60 bu. oats, 1 acre rye, 10 bu. rye, 2 acres wheat, 50 bu. wheat, 1/2 acre Irish potatoes, 25 bu. Irish potatoes.
- 1/4 acre with 10 apple trees; 10 peach trees.
- 5 cords of wood cut valued at $5.

*Federal census lists William, 49; Lizzie, 40 (wife); Eliza, 5; John, 3; Lydia, 2; William, 6/12 (Dec.).*

## George Trostle

- Owner
- 40 tilled acres, 2 acres permanent meadows, 2 acres woodland forests, 2 acres unimproved land, valued at $2500.
- Farm machinery valued at $200; livestock, $200; fences, $25; fertilizers, $50; wages paid for farm labor, $25; weeks of hired labor, 4; estimated value of all farm productions, $400.
- Acres of grass mown, 6; not mown, 6; 6 tons of hay harvested.
- 1 horse, 3 milch cows, 3 other cattle, 3 calves dropped, 1 cattle slaughtered, 300 lbs. butter, 4 swine, 15 poultry in barnyard, 100 dozen eggs produced.
- 6 acres Indian corn, 100 bu. Indian corn, 6 acres oats, 150 bu. oats, 1 acre rye, 10 bu. rye, 6 acres wheat, 100 bu. wheat, 1/2 acre Irish potatoes, 40 bu. Irish potatoes.
- 1/2 acre with 10 apple trees; 10 peach trees.
- 5 cords of wood cut valued at $5.

*Federal census lists George, 67; Cristine, 60 (wife); Emmaline, 18.*

## Otto Roland

- Owner
- 15 tilled acres, 2 acres permanent meadows, 1 acre unimproved land, valued at $2000.
- Farm machinery valued at $300; livestock, $200; fences, $25; fertilizers, $20; estimated value of all farm productions, $150.
- Acres of grass mown, 4; 4 tons of hay harvested.
- 1 horse, 2 milch cows, 1 other cattle, 2 calves dropped, 1 cattle slaughtered, 200 lbs. butter, 3 swine, 50 poultry in barnyard, 300 dozen eggs produced.
- 2 acres Indian corn, 60 bu. Indian corn, 1 acre oats, 55 bu. oats, 1/4 acre rye, 5 bu. rye, 2 acres wheat, 35 bu. wheat, 1/2 acre Irish potatoes, 25 bu. Irish potatoes.
- 1/2 acre Irish potatoes, 25 bu. Irish potatoes.
- 1/4 acre with 20 apple trees; 20 peach trees, 4 bu. peaches; total value of orchard products, $4.

*Federal census lists Otto, 43 (widower), a taylor; Otto, 19, farm laborer; Augustus, 17, farm laborer; Emma, 11.*

## Andrew Emmert

- Owner
- 40 tilled acres, 6 acres permanent meadows, 3 acres woodland forests, 4 acres unimproved land, valued at $4000.
- Farm machinery valued at $500; livestock, $500; fences, $150; fertilizers, $75; wages paid for farm labor, $300; weeks of hired labor, 50; estimated value of all farm productions, $1000.
- Acres of grass mown, 14; not mown, 9; 20 tons of hay harvested.
- 3 horses, 6 milch cows, 6 other cattle, 6 calves dropped, 1 cattle sold living, 1 cattle slaughtered, 500 lbs. of butter, 5 swine, 75 poultry in barnyard, 400 dozen eggs produced.
- 12 acres Indian corn, 300 bu. Indian corn, 5 acres oats, 171 bu. oats, 8 acres wheat, 130 bu. wheat, 1 acre Irish potatoes, 100 bu. Irish potatoes, 1/5 acre sweet potatoes, 16 bu. sweet potatoes, 1 1/2 acres tobacco, 2500 lbs. tobacco.
- 1/2 acre with 26 apple trees, 10 bu. apples, 6 peach trees, 1 bu. peaches; total value of orchard products, $5.
- Market gardens valued at $10.
- 15 cords of wood cut valued at $15.

*Federal census lists Andrew, 53, a miller; Anna M., 49 (wife); Walter J., 23; Franklin B. H., 20; Ida H., 17; Milton H., 15; Amelia H., 13; Clara K., 11; Mary A., 9; Ellen, 8.*

## John Harnley

- Owner
- 58 tilled acres, 6 acres permanent meadows, 2 acres woodland forests, 5 acres unimproved land, valued at $2500.
- Farm machinery valued at $600; livestock, $600; fences,, $100; fertilizers, $50; wages paid for farm labor, $160; weeks of hired labor, 25; estimated value of all farm productions, $500.

- Acres of grass mown, 15; not mown, 10; 10 tons of hay harvested.
- 3 horses, 4 milch cows, 3 other cattle, 4 calves dropped, 2 cattle sold living, 2 cattle slaughtered, 400 lbs. butter, 4 swine, 50 poultry in barnyard, 300 dozen eggs produced.
- 14 acres Indian corn, 300 bu. Indian corn, 7 acres oats, 325 bu. oats, 9 acres rye, 40 bu. rye, 7 acres wheat, 75 bu. wheat, 1/4 acre Irish potatoes, 20 bu. Irish potatoes.
- 1/2 acre with 60 apple trees; 1/4 acre with 30 peach trees.
- 1 cord of wood cut valued at $1.

*Federal census lists John, 35, a brush maker; Louisa, 30 (wife); Frank, 13; Charles, 9; John, 7; Albert, 6; Emma, 4; Louisa, 2; William, 4/12 (Mar.).*

### Henry Trostle

- Owner
- 48 tilled acres, 6 acres permanent meadows, 1 acre woodland forests, 15 acres unimproved land, valued at $3000.
- Farm machinery valued at $500; livestock, $300; fences, $100; fertilizers, $100; wages paid for farm labor, $125; weeks of hired labor, 20; estimated value of all farm productions, $500.
- Acres of grass mown, 8; not mown, 3; 15 tons of hay harvested.
- 1 horse, 5 milch cows, 2 other cattle, 5 calves dropped, 400 lbs. butter, 30 poultry in barnyard, 200 dozen eggs produced.
- 4 acres Indian corn, 50 bu. Indian corn, 4 acres oats, 125 bu. oats, 4 acres wheat, 53 bu. wheat, 1/4 acre Irish potatoes, 20 bu. Irish potatoes.
- 1/2 acre with 12 apple trees.
- 5 cords of wood cut valued at $5.

*Federal census lists Henry, 62 (single); Catherine, 48 (sister); Benjamin Good, 22 (farm laborer).*

### Benjamin Behler

- Owner
- 9 tilled acres, valued at $1000.
- Farm machinery valued at $200; livestock, $200; fences, $100; fertilizers, $50; wages paid for farm labor, $25; weeks of hired labor, 4; estimated value of all farm productions, $310.
- Acres of grass mown, 4; 4 tons of hay harvested.
- 1 milch cow, 1 calf dropped, 150 lbs. butter, 2 swine, 30 poultry in barnyard, 200 dozen eggs produced.
- 2 acres Indian corn, 50 bu. Indian corn, 2 acres oats, 50 bu. oats, 1 acre rye, 20 bu. rye, 1 acre Irish potatoes, 40 bu. Irish potatoes.
- 1/16 acre with 10 apple trees, 30 bu. apples; 2 acres with 20 peach trees, 8 bu. peaches; total value of orchard products, $20.
- 6 cords of wood cut valued at $6.

*Federal census lists Benjamin, 38; Catherine, 39 (wife); John, 14; Sarah, 13.*

### Jacob Gross

- Owner
- 40 tilled acres, 5 acres permanent meadows, 1 acre woodland forests, 10 acres unimproved land, valued at $3100.
- Farm machinery valued at $500; livestock, $400; fences, $100; fertilizers, $50; wages paid for farm labor, $25; weeks of hired labor, 4; estimated value of all farm productions, $300.
- Acres of grass mown, 6; 6 tons of hay harvested.
- 2 horses, 2 milch cows, 5 other cattle, 2 calves dropped, 200 lbs. butter, 2 swine, 12 poultry in barnyard, 2 other poultry, 100 dozen eggs produced.
- 8 acres Indian corn, 100 bu. Indian corn, 8 acres oats, 100 bu. oats, 1 acre rye, 20 bu. rye, 10 acres wheat, 100 bu. wheat, 1/2 acre Irish potatoes, 20 bu. Irish potatoes.
- 1/4 acre with 25 apple trees, 25 bu. apples; 2 acres with 5 peach trees; total value of orchard products, $10.
- 5 cords wood cut valued at $5.

*Federal census lists Jacob, 49; Wilhelmina, 52 (wife); Charles, 25; Wilhelmina, 18; Frederic, 16; Annie, 9; John, 8 (grandson).*

### Amandus Gangaway

- Owner
- 40 tilled acres, 2 acres permanent meadows, 1 acre woodland forests, 10 acres unimproved land, valued at $2000.
- Farm machinery valued at $100; livestock, $500; fences, $50; fertilizers, $50; wages paid for farm labor, $25; weeks of hired labor, 4; estimated value of all farm productions, $400.
- Acres of grass mown, 6; not mown, 4; 6 tons of hay harvested.
- 3 horses, 4 milch cows, 5 other cattle, 4 calves dropped, 400 lbs. butter, 1 swine, 50 poultry in barnyard, 200 dozen eggs produced.

- 5 acres Indian corn, 100 bu. Indian corn, 5 acres oats, 100 bu. oats, 1 acre rye, 10 bu. rye, 5 acres wheat, 110 bu. wheat, 1 acres Irish potatoes, 40 bu. Irish potatoes.
- 1/4 acre with 25 apple trees, 10 bu. apples; 2 acres with 10 peach trees; total value of orchard products, $10.
- 5 cords of wood cut valued at $5.

*Federal census lists Amandus, 40; Rachel, 35 (wife); Israel, 7; Susanna, 2; Jonathan, 10/12 (Aug.); Joshua, 33 (brother).*

### John Stover

- Rents for shares of products.
- 30 tilled acres, 2 acres permanent meadows, 1 acre woodland forests, 10 acres unimproved land, valued at $1500.
- Farm machinery valued at $100; livestock, $200; fences, $25; fertilizers, $25; wages paid for farm labor, $25; weeks of hired labor, 4; estimated value of all farm productions, $300.
- Acres of grass mown, 5; not mown, 5; 10 tons of hay harvested.
- 2 milch cows, 5 other cattle, 2 calves dropped, 200 lbs. butter, 2 swine, 25 poultry in barnyard, 100 dozen eggs produced.
- 4 acres Indian corn, 100 bu. Indian corn, 4 acres oats, 75 bu. oats, 1 acre rye, 10 bu. rye, 2 acres wheat, 50 bu. wheat, 1 acre Irish potatoes, 40 bu. Irish potatoes.
- 1/4 acre with 25 apple trees, 10 bu. apples; 2 acres with 10 peach trees; total value of orchard products, $10.
- 5 cords of wood cut valued at $5.

*Federal census lists John, 35; Sally, 30 (wife); Eva, 4; Henry, 3; baby daughter, 4/12 (Feb.); Mary Redcay, 47 (mother-in-law).*

### Samuel Good

- Rents for shares of products.
- 75 tilled acres, 6 acres permanent meadows, 2 acres woodland forests, 15 acres unimproved land, valued at $3000.
- Farm machinery valued at $600; livestock, $600; fences, $50; fertilizers, $100; wages paid for farm labor, $150; weeks of hired labor, 25; estimated value of all farm productions, $1100.
- Acres of grass mown, 30; not mown, 5; 25 tons of hay harvested.
- 5 horses, 2 mules and asses, 6 milch cows, 11 other cattle, 6 calves dropped, 3 cattle sold living, 1 cattle slaughtered, 600 lbs. butter, 4 swine, 50 poultry in barnyard, 500 dozen eggs produced.
- 13 acres Indian corn, 600 bu. Indian corn, 16 acres oats, 550 bu. oats, 14 acres wheat, 225 bu. wheat, 1/4 acre Irish potatoes, 25 bu. Irish potatoes.
- 1/2 acre with 30 apple trees, 5 bu. apples; total value of orchard products, $2.
- Market gardens valued at $1.
- 10 cords of wood cut valued at $10.

*Federal census lists Samuel, 37; Sarah, 29 (wife); Ida, 10; James, 7; Wayne, 5; Elmina, 3; Sarah A., 1; Henry Wolf, 32, farm laborer.*

### Joseph Zerbe

- Owner
- 40 tilled acres, 5 acres permanent meadows, 1 acre woodland forests, 60 acres unimproved land, valued at $2000.
- Farm machinery valued at $600; livestock, $600; fences, $50; fertilizers, $100; wages paid for farm labor, $150; weeks of hired labor, 25; estimated value of all farm productions, $700.
- Acres of grass mown, 25; not mown, 6; 20 tons of hay harvested.
- 2 horses, 8 milch cows, 5 other cattle, 8 calves dropped, 1 cattle slaughtered, 600 lbs. butter, 5 swine, 21 poultry in barnyard, 100 dozen eggs produced.
- 10 acres Indian corn, 200 bu. Indian corn, 10 acres oats, 100 bu. oats, 1 acre rye, 10 bu. rye, 10 acres wheat, 200 bu. wheat, 1/2 acre Irish potatoes, 20 bu. Irish potatoes.
- 1 acre with 25 apple trees, 5 bu. apples, 2 acres with 10 peach trees; total value of orchard products, $2.
- 10 cords of wood cut valued at $10.

*Federal census lists Joseph, 56; Susanna, 61 (wife); Joseph, 20; Susanna, 17; Samuel, 15.*

### Isaac Steffe

- Owner
- 50 tilled acres, 6 acres woodland forests, 4 acres unimproved land, valued at $2000.
- Farm machinery valued at $500; livestock, $500; fences, $100; fertilizers, $50; wages paid for farm labor, $100; weeks of hired labor, 20; estimated value of all farm productions, $300.
- Acres of grass mown, 10; not mown, 10; 5 tons of hay harvested.
- 2 horses, 2 milch cows, 1 other cattle, 2 calves dropped, 200 lbs. butter, 36 poultry in barnyard, 100 dozen eggs produced.
- 12 acres Indian corn, 200 bu. Indian corn, 6 acres oats, 76 bu. oats, 2 acres rye, 8 bu. rye, 2 acres wheat, 13 bu. wheat, 1 acre Irish potatoes, 140 bu. Irish potatoes.
- 1 acre with 100 apple trees, 20 bu. apples; 2 acres with 25 peach trees; total value of orchard products, $25.

• Market gardens valued at $50.
• 10 cords of wood cut valued at $10.

*Federal census lists Isaac, 37; Elizabeth, 37 (wife); Nathaniel Gangaway, 13 (farm laborer); Christian Good, 82; Susan Good, 75; Lydia Good (sister), 34.*

## William Steffe

• Owner
• 28 tilled acres, 2 acres permanent meadows, valued at $2700.
• Farm machinery valued at $100; livestock, $300; fences, $25, fertilizers, $25; wages paid for farm labor, $15; weeks of hired labor, 2; estimated value of all farm productions, $300.
• Acres of grass mown, 10; not mown, 10; 10 tons of hay harvested.
• 2 horses, 4 milch cows, 1 other cattle, 4 calves dropped, 1 cattle slaughtered, 1 cattle died, strayed, stolen, and not recovered, 400 lbs. butter, 5 swine, 50 poultry in barnyard, 300 dozen eggs produced.
• 4 acres Indian corn, 100 bu. Indian corn, 4 acres oats, 122 bu. oats, 5 acres wheat, 60 bu. wheat, 1 acre Irish potatoes, 40 bu. Irish potatoes.
• 1/4 acre with 10 apple trees, 20 bu. apples; total value of orchard products, $10.
• Market gardens valued at $10.
• 10 cords of wood cut valued at $10.

*Federal census lists William, 38; Emma, 27 (wife); Augustus, 7; Levi, 68 (father); Rachel, 69 (mother).*

## Jacob Leinbach

• Owner
• 20 tilled acres, 4 acres permanent meadows, 1 acre unimproved land, valued at $1300.
• Farm machinery valued at $200; livestock, $200; fences, $25; fertilizers, $16; wages paid for farm labor, $10; weeks of hired labor, 2; estimated value of all farm productions, $400.
• Acres of grass mown, 10; 15 tons of hay harvested.
• 1 horse, 2 milch cows, 2 other cattle, 2 calves dropped, 1 cattle slaughtered, 200 lbs. butter, 3 swine, 40 poultry in barnyard, 300 dozen eggs produced.
• 3 acres Indian corn, 50 bu. Indian corn, 3 acres oats, 100 bu. oats, 1 acre rye, 23 bu. rye, 2 acres wheat, 35 bu. wheat, 1/4 acre Irish potatoes, 20 bu. Irish potatoes.
• 1/2 acre with 15 apple trees, 10 bu. apples; 2 acres with 6 peach trees; total value of orchard products, $5.

*Federal census lists Jacob, 58; Susannah, 54 (wife); Jacob, 19.*

## Noah Gehman

• Owner
• 30 tilled acres, 4 acres permanent meadows, 6 acres woodland forests, 1 acre unimproved land, valued at $3000.
• Farm machinery valued at $600; livestock, 600; fences, $60; fertilizers, $50; wages paid for farm labor, $50; weeks of hired labor, 8; estimated value of all farm productions, $500.
• Acres of grass mown, 10; not mown, $10; 15 tons of hay harvested.
• 2 horses, 5 milch cows, 2 other cattle, 5 calves dropped, 1 cattle died, strayed, stolen, and not recovered, 4 swine, 50 poultry in barnyard, 400 dozen eggs produced.
• 6 acres Indian corn, 200 bu. Indian corn, 6 acres oats, 200 bu. oats, 1 acre rye, 10 bu. rye, 6 acres wheat, 120 bu. wheat, 1/2 acre Irish potatoes, 50 bu. Irish potatoes, 1/2 acre tobacco, 400 lbs. tobacco.
• 12 apple trees.
• Market gardens valued at $5.
• 10 cords of wood cut valued at $10.

*Federal census lists Noah, 23; Hannah, 23 (wife); David, 11; Benjamin, 72 (father); Elizabeth, 67 (mother).*

## Aaron Good

• Owner
• 26 tilled acres, valued at $2000.
• Farm machinery valued at $400; livestock, $400; fences, $50; fertilizers, $25; wages paid for farm labor, $25; weeks of hired labor, 6; estimated cost of all farm productions, $400.
• Acres of grass mown, 4; 6 tons of hay harvested.
• 1 horse, 2 milch cows, 1 other cattle, 2 calves dropped, 200 lbs. butter, 2 swine, 12 poultry in barnyard, 100 dozen eggs produced.
• 5 acres Indian corn, 100 bu. Indian corn, 5 acres oats, 100 bu. oats, 1 acre rye, 10 bu. rye, 5 acres wheat, 100 bu. wheat, 1/2 acre Irish potatoes, 25 bu. Irish potatoes.
• 1/4 acre with 20 apple trees, 5 bu. apples; 1/4 acre with 20 peach trees, 5 bu. peaches; total value of orchard products, $5.
• 5 cords of wood cut valued at $5.

*Federal census lists Aaron, 28; Betsey, 26 (wife); Mary, 5.*

### Joseph Horning

- Owner
- 15 tilled acres, 2 acres permanent meadows, 3 acres woodland forests, 6 acres unimproved land, valued at $1800.
- Farm machinery valued at $400; livestock, $200; fences, $100; fertilizers, $25; wages paid for farm labor, $25; weeks of hired labor, 4; estimated value of all farm productions, $300.
- Acres of grass mown, 3, 4 tons of hay harvested.
- 1 horse, 2 milch cows, 2 calves dropped, 1 cattle slaughtered, 100 lbs. butter, 3 swine, 30 poultry in barnyard, 100 dozen eggs produced.
- 2 acres Indian corn, 60 bu. Indian corn, 2 acres oats, 75 bu. oats, 1 acre rye, 15 bu. rye, 2 acres wheat, 32 bu. wheat, 1/4 acre Irish potatoes, 25 bu. Irish potatoes.
- 1/4 acre with 20 apple trees, 5 bu. apples; 5 peach trees; total value of orchard products, $4.
- Market gardens valued at $1.
- 5 cords of wood cut valued at $5.

*Federal census lists Joseph, 29; Lydia, 29 (wife); Hettie, 5; Mary, 1; Mary Gehman, 55 (Lydia's aunt).*

### Daniel Brendle

- Owner
- 25 tilled acres, 2 acres permanent meadows, 2 acres woodland forests, 10 acres unimproved land, valued at $1800.
- Farm machinery valued at $300, livestock, $200; fences, $25; fertilizers, $25; wages paid for farm labor, $10; weeks of hired labor, 2; estimated value of all farm productions, $250.
- Acres of grass mown, 6; not mown, 6; 6 tons of hay harvested.
- 2 horses, 4 milch cows, 2 other cattle, 4 calves dropped, 1 cattle slaughtered, 300 lbs. butter, 5 swine, 15 poultry in barnyard, 10 other poultry, 100 dozen eggs produced.
- 2 acres Indian corn, 20 bu. Indian corn, 2 acres oats, 25 bu. oats, 2 acres wheat, 25 bu. wheat, 1/2 acre Irish potatoes, 25 bu. Irish potatoes.
- 1/4 acre with 7 apple trees, 1/2 acre with 7 peach trees; total value of orchard products, $2.
- Market gardens valued at $1.
- 10 cords of wood cut valued at $10.

*Federal census lists Daniel, 62; Susan, 59 (wife).*

### William Echternacht

- Owner
- 150 tilled acres, 10 acres permanent meadows, 10 acres woodland forests, 10 acres unimproved land, valued at $10,000.
- Farm machinery valued at $600; livestock, $600; fences, $100; fertilizers, $100; wages paid for farm labor, $150; weeks of hired labor, 150; estimated value of all farm productions, $1800.
- Acres of grass mown, 20; not mown, 10; 25 tons of hay harvested.
- 7 horses, 10 milch cows, 17 other cattle, 10 calves dropped, 1 cattle slaughtered, 600 lbs. butter, 8 swine, 100 poultry in barnyard, 510 dozen eggs produced.
- 15 acres Indian corn, 600 bu. Indian corn, 15 acres oats, 600 bu. oats, 5 acres rye, 100 bu. rye, 15 acres wheat, 300 bu. wheat, 1/2 acre Irish potatoes, 75 bu. Irish potatoes.
- 1 acre with 50 apple trees, 10 peach trees; total value of orchard products, $1.
- 10 cords of wood cut valued at $10.

*Federal census lists William, 54; Caroline, 49 (wife); Albert, 22; Justus, 21; Mary, 19; Emma, 17; Henry, 13; George, 10.*

### Benjamin Slabach

- Owner
- 35 tilled acres, 5 acres permanent meadows, 2 acres woodland forests, 8 acres unimproved land, valued at $3000.
- Farm machinery valued at $400; livestock, $400; fences, $100; fertilizers, $50; wages paid for farm labor, $50; hours of hired labor, $50; estimated value of all farm productions, $450.
- Acres of grass mown, 10; not mown, 10; 15 tons of hay harvested.
- 5 horses, 6 milch cows, 6 calves dropped, 1 cattle slaughtered, 600 lbs. butter, 3 swine, 75 poultry in barnyard, 200 dozen eggs produced.
- 5 acres Indian corn, 100 bu. Indian corn, 5 acres oats, 100 bu. oats, 5 acres rye, 20 bu. rye, 6 acres wheat, 100 bu. wheat, 1 acre Irish potatoes, 40 bu. Irish potatoes.
- 1 acre with 25 apple trees, 10 peach trees; total value of orchard products, $1.
- 10 cords of wood cut valued at $5.

*Federal census lists Benjamin, 57; Matty, 57 (wife); Lydia, 21; John, 18.*

### Henry Gehman

- Rents for shares of products.
- 60 tilled acres, 6 acres permanent meadows, 20 acres woodland forests, 20 acres unimproved land, valued at $4000.

- Farm machinery valued at $500; livestock, $600; fences, $50; fertilizers, $50; wages paid for farm labor, $100; weeks of hired labor, 20; estimated value of all farm productions, $600.
- Acres of grass mown, 20; not mown, 10; 25 tons of hay harvested.
- 4 horses, 6 milch cows, 6 other cattle, 6 calves dropped, 1 cattle slaughtered, 300 lbs. butter, 6 swine, 100 poultry in barnyard, 500 dozen eggs produced.
- 8 acres Indian corn, 175 bu. Indian corn, 8 acres oats, 200 bu. oats, 1 acre rye, 10 bu. rye, 8 acres wheat, 100 bu. wheat, 1/2 acre Irish potatoes, 25 bu. Irish potatoes.
- 1 acre with 20 apple trees, 1 acre peach trees.
- 12 cords of wood cut valued at $12.

*Federal census lists Henry, 35; Fanny, 37 (wife); Joseph, 18; Anna, 8; Hattie, 7; Joseph, 70, and Esther, 65, next house (possibly a grandparent home).*

## Henry Keiffer

- Owner
- 70 tilled acres, 5 acres permanent meadows, valued at $3000.
- Farm machinery valued at $400; livestock, $400; fences, $50; fertilizers, $50; wages paid for farm labor, $50; weeks of hired labor, 6; estimated value of all farm productions, $600.
- Acres of grass mown, 15; not mown, 10; 15 tons of hay harvested.
- 4 horses, 3 milch cows, 5 other cattle, 3 calves dropped, 1 cattle slaughtered, 300 lbs. butter, 5 swine, 30 poultry in barnyard, 200 dozen eggs produced.
- 10 acres Indian corn, 250 bu. Indian corn, 10 acres oats, 300 bu. oats, 1 acre rye, 20 bu. rye, 9 acres wheat, 80 bu. wheat, 1/2 acre Irish potatoes, 50 bu. Irish potatoes.
- 1/2 acre with 20 apple trees, 12 peach trees, 5 bu. peaches; total value of orchard products, $5.
- 10 cords of wood cut valued at $10.

*Federal census lists Henry, 39; Sybilla, 33 (wife); Maggie, 5; Mary A., 3.*

## Edward Stoner

- Owner
- 40 tilled acres, 5 acres permanent meadows, 1 acre woodland forests, 5 acres unimproved land, valued at $2000.
- Farm machinery valued at $400; livestock, $400; fences, $56; fertilizers, $50; wages paid for farm labor, $25; weeks of hired labor, 4; estimated value of all farm productions, $700.
- Acres of grass mown, 15; not mown, 10; 20 tons of hay harvested.
- 3 horses, 6 milch cows, 6 calves dropped, 720 lbs. butter, 4 swine, 40 poultry in barnyard, 600 dozen eggs produced.
- 8 acres Indian corn, 200 bu. Indian corn, 8 acres oats, 300 bu. oats, 1 acre rye, 26 bu. rye, 4 acres wheat, 60 bu. wheat, 1 acre Irish potatoes, 60 bu. Irish potatoes.
- 1 acre with 12 apple trees, 25 peach trees, 6 bu. peaches; total value of orchard products, $6
- Market gardens valued at $20
- 10 cords of wood cut valued at $10.

*Federal census lists Edward, 37; Catherine, 33 (wife).*

## Levi Trostle

- Owner
- 20 tilled acres, 1 acre permanent meadows, 1 acre woodland forests, 1 acre unimproved land, valued at $1000.
- Farm machinery valued at $100; livestock, $200; fences, $25; fertilizers, $25; wages paid for farm labor, $12; weeks of hired labor, 2; estimated value of all farm productions, $300.
- Acres of grass mown, 5; not mown, 5; 10 tons of hay harvested.
- 1 horse, 4 milch cows, 4 calves dropped, 400 lbs. butter, 4 swine, 20 poultry in barnyard, 300 dozen eggs produced.
- 2 acres Indian corn, 100 bu. Indian corn, 2 acres oats, 100 bu. oats, 1 acre rye, 25 bu. rye, 2 acres wheat, 50 bu. wheat, 1 acre Irish potatoes, 25 bu. Irish potatoes.
- 1 acre with 10 apple trees, 10 peach trees, total value of orchard products, $1.
- 5 cords of wood cut valued at $5.

*Federal census lists Levi, 45; Anna, 50 (wife); Levi, 10; William, 28; Ellmar, 18.*

## Solomon Heft

- Owner
- 20 tilled acres, 1 acre permanent meadows, 1 acre woodland forests, 1 acre unimproved land, valued at $1000.
- Farm machinery valued at $100; livestock, $100; fences, $25; fertilizers, $25; wages paid for farm labor, $10; weeks of hired labor, 2; estimated value of all farm productions, $400.
- Acres of grass mown, 5; not mown, 1; 8 tons of hay harvested.
- 2 horses, 4 milch cows, 2 other cattle, 4 calves dropped, 1 cattle slaughtered, 400 lbs. butter, 4 swine, 20 poultry in barnyard, 200 dozen eggs produced.
- 5 acres Indian corn, 100 bu. Indian corn, 2 acres oats, 50 bu. oats, 1 acre rye, 30 bu. rye, 2 acres wheat, 50 bu. wheat, 1 acre Irish potatoes, 50 bu. Irish potatoes.

• 1 acre with 10 apple trees, 10 peach trees; total value of orchard products, $1.
• 4 cords of wood cut valued at $10.
*Federal census lists Solomon, 41; Eliza, 41 (wife).*

### William Sweitzer
• Rents for shares of products.
• 50 tilled acres, 9 acres permanent meadows, 3 acres woodland forests, 11 acres unimproved land, valued at $3000.
• Farm machinery valued at $300; livestock, $500; fences, $100; fertilizers, $70; wages paid for farm labor, $30; weeks of hired labor, 10; estimated value of all farm productions, $600.
• Acres of grass mown, 30; not mown, 10; 20 tons of hay harvested.
• 2 horses, 5 milch cows, 9 other cattle, 5 calves dropped, 500 lbs. butter, 4 swine, 50 poultry in barnyard, 300 dozen eggs produced.
• 7 acres Indian corn, 250 bu. Indian corn, 10 acres oats, 272 bu. oats, 10 acres wheat, 110 bu. wheat, 1/2 acre Irish potatoes, 20 bu. Irish potatoes.
• 2 acres with 6 apple trees, 2 bu. apples; 1/16 acre with 6 peach trees; total value of orchard products, $1.
• 10 cords of wood cut valued at $11.
*Federal census lists William, 32; Wilmina, 39 (wife); Lily Ann, 9; Rose Ellen, 7; Emma C., 6; Peter O., 4; David L., 2; Sybilla W., 1; Sarah A., 3/12 (Mar.).*

### Esaias Brendle
• Owner
• 40 tilled acres, 2 acres permanent meadows, 3 acres woodland forests, 10 acres unimproved land, valued at $3500.
• Farm machinery valued at $300; livestock, $500; fences, $100; fertilizers, $10; wages paid for farm labor, $30; weeks of hired labor, 10; estimated value of all farm productions, $400.
• Acres of grass mown, 14; not mown, 10; 10 tons of hay harvested.
• 2 horses, 5 milch cows, 4 other cattle, 1 cattle died, strayed, stolen, and not recovered, 500 lbs. butter, 40 poultry in barnyard, 300 dozen eggs produced.
• 12 acres Indian corn, 200 bu. Indian corn, 10 acres oats, 200 bu. oats, 10 acres wheat, 110 bu. wheat, 2 acres Irish potatoes, 150 bu. Irish potatoes.
• 1/2 acre with 30 apple trees, 5 bu. apples; 1/16 acre with 25 peach trees; 4 bu. peaches, total value of orchard products, $10.
• Market gardens valued at $20.
• 10 cords of wood cut valued at $10.
*Federal census lists Esaias, 37; Elisa, 30 (wife); Warren, 12; Mary E., 8; Ida C., 6; Willie H., 5; Esaias F., 4; Cyrus H., 2.*

### Jno. M. Styer
• Owner
• 30 tilled acres, 8 acres permanent meadows, 300 acres woodland forests, 20 acres unimproved land, valued at $3500.
• Farm machinery valued at $300; livestock, $500; fences, $100; fertilizers, $100; wages paid for farm labor, $150; weeks of hired labor, 25; estimated value of all farm productions, $700.
• Acres of grass mown, 12; not mown, 300; 13 tons of hay harvested.
• 2 horses, 6 milch cows, 4 other cattle, 6 calves dropped, 1 cattle slaughtered, 2 cattle died, strayed, stolen, and not recovered, 600 lbs. butter, 3 swine, 50 poultry in barnyard, 500 dozen eggs produced.
• 5 acres Indian corn, 200 bu. Indian corn, 2 acres oats, 20 bu. oats, 2 acres rye, 25 bu. rye, 3 acres wheat, 50 bu. wheat, 3/4 bu. Irish potatoes, 50 bu. Irish potatoes, 6 bu. sweet potatoes, 1/2 acre tobacco, 1300 lbs. tobacco.
• 1/2 acre with 20 apple trees, 5 bu. apples; 25 peach trees, 10 bu. peaches; total value of orchard products, $10.
• 10 cords of wood cut valued at $10.
*Federal census lists Martin, 21, farmer; Mary, 46 (mother); D. Franklin, 19 (son), farm laborer; Annie M., 17 (daughter); Lizzie, 6 (daughter).*

### John Sparr
• Rents for shares of products
• 25 tilled acres, 2 acres permanent meadows, 5 acres woodland forests, 5 acres unimproved land, valued at $800.
• Farm machinery valued at $400; livestock, $400; fences, $50; fertilizers, $50; estimated value of all farm productions, $500.
• Acres of grass mown, 4; not mown, 10; 5 tons of hay harvested.
• 1 horse, 4 milch cows, 1 other cattle, 1 cattle slaughtered, 400 lbs. butter, 3 swine, 50 poultry in barnyard, 200 dozen eggs produced.
• 5 acres Indian corn, 125 bu. Indian corn, 2 acres Irish potatoes, 250 bu. Irish potatoes, 2/3 acre tobacco, 1300 lbs. tobacco.

- 1/4 acre with 12 apple trees, 40 bu. apples; total value of orchard products, $20.
- Market gardens valued at $150.
- 10 cords of wood cut valued at $10.

*Federal census lists John, 46; Sarah, 35; John, 18; Hattie, 16; Sarah, 14; William, 12; Catharina, 11; Louisa, 7; Henry, 4; Clara, 1/12 (Apr.).*

## William Sweitser

- Rents for shares of products.
- 50 tilled acres, 2 acres permanent meadows, 5 acres woodland forests, 10 acres unimproved land, valued at $1,000.
- Farm machinery valued at $400; livestock, $400; fences, $50; fertilizers, $50; wages paid for farm labor, $50; weeks of hired labor, 12; estimated value of all farm productions, $500.
- Acres of grass mown, 10; not mown, 10; 10 tons of hay harvested.
- 2 horses, 2 milch cows, 2 calves dropped, 1 cattle slaughtered, 200 lbs. butter, 100 poultry in barnyard, 400 dozen eggs produced.
- 2 acres Indian corn, 50 bu. Indian corn, 2 acres oats, 100 bu. oats, 1 acre rye, 10 bu. rye, 2 acres wheat, 50 bu. wheat, 2 acres Irish potatoes, 100 bu. Irish potatoes.
- 1/2 acre with 20 apple trees, 11 bu. apples.
- 10 cords of wood cut valued at $10.

*Federal census lists William, 40; Sophia, 30 (wife); Hannah, 12; Ada, 10; Annie, 8; Pierce F., 5; Peter E., 4; baby son, 5/12 (Jan.).*

## Franklin S. Stover

- Rents for shares of products.
- 15 tilled acres, 1 acre permanent meadows, valued at $1,000.
- Farm machinery valued at $100; livestock, $200; fences, $25; fertilizers, $50; wages paid for farm labor, $25; weeks of hired labor, 4; estimated value of all farm productions, $300.
- Acres of grass mown, 3; not mown, 4; 5 tons of hay harvested.
- 3 milch cows, 2 other cattle, 3 calves dropped, 1 cattle slaughtered, 400 lbs. butter, 3 swine, 25 poultry in barnyard, 200 dozen eggs produced.
- 2 acres Indian corn, 100 bu. Indian corn, 2 acres oats, 70 bu. oats, 2 acres wheat, 43 bu. wheat, 1/4 acre Irish potatoes, 20 bu. Irish potatoes.
- 20 apple trees, 5 bu. apples, 1/2 acre with 4 peach trees, 1 bu. peaches; total value of orchard products, $1.

*Federal census lists Franklin, 25; Amanda E., 22 (wife); Lizzie Agnes, 3; Catherine Stover, 69 (aunt).*

# Index

— C —



— D —

— H —

— L —



— M —

— N —



— O —



— P —

— T —

— Y —



— Z —